Carlton Foster and Co.

Wholesale Manufacturers of Sash, Doors, Blinds, Mouldings, Brackets etc.

Stair Work, Newels and Inside Finish

Carlton Foster and Co.

Wholesale Manufacturers of Sash, Doors, Blinds, Mouldings, Brackets etc.
Stair Work, Newels and Inside Finish

ISBN/EAN: 9783744650977

Printed in Europe, USA, Canada, Australia, Japan

Cover: Foto ©berggeist007 / pixelio.de

More available books at **www.hansebooks.com**

1888.

CARLTON FOSTER & CO.,

Wholesale Manufacturers of

SASH,

DOORS,

BLINDS,

Mouldings, Brackets, &c.

Cor. 22nd and Union Sts.,
CHICAGO, ILL.

John Hicks, Printer, Oshkosh.

PRICE LIST.

MARCH 1, 1888.

CARLTON FOSTER & CO.,

WHOLESALE MANUFACTURERS OF

Sash, Doors, Blinds,

MOULDINGS, BRACKETS, ETC.,

Stair Work, Newels and Inside Finish.

WAREHOUSE AND OFFICE,
Cor. West 22d and Union Sts.
CHICAGO, ILL.

MILLS AND OFFICE,
Cor. Oregon and Sixth Sts.
OSHKOSH, WIS.

John Hicks, Printer, Oshkosh, Wis.

OSHKOSH, WIS., }
CHICAGO, ILL., } March 15, 1888.

To Our Patrons:

Herewith you will find our 1888 List containing price-lists of Sash, Doors, Blinds, Frames, Stairwork, Fancy Glass, Brackets and Cottage Front Doors.

We desire to call your attention to our popular line of Cottage Front Doors, a few styles of which are shown on pages 42 to 50. We make these doors with a 12½ *inch Lock Rail.*

We make all styles of Sash, Doors and Blinds used in any market of the country, as well as numerous specialties, such as Ventilators, Mantels, Pew Ends, Veneered Doors, Store Fronts, Stairwork, Brackets, Church Sash, Queen Anne Sash, etc., etc.

The quality of our goods is well known to the trade, and we guarantee to keep them up to their usual high grade.

We carry in our Chicago warehouses a very complete stock, which was manufactured by ourselves at our factory in Oshkosh.

We invite a personal inspection of our stock and hope to be favored with your orders.

Estimates, Price Lists and Moulding Books furnished upon application.

Very Truly,

CARLTON FOSTER & CO.

GRADES OF

Sash, Doors and Blinds,

ADOPTED BY THE

Wholesale Sash, Door and Blind Manufacturers' Association of the Northwest, November 16th, 1887.

DOORS.

No. 1 Doors—Workmanship on No. 1 doors must be good. Stiles, rails and panels must be clear, except that white sap and water stain caused by cross piling lumber, is admitted, and small pin knots not exceeding one-fourth (¼) inch in diameter may be allowed. The Standard No. 1 door shall be pinned and the wedges glued.

No. 2 Doors—No. 2 doors may contain knots not larger than one (1) inch in diameter, and may contain blue sap on two (2) sides, and may contain gum spots showing on one (1) side. Other small defects may be allowed, but the total number of defects, (not including blue sap) shall not exceed ten (10) in number on each side, and blue sap must not exceed fifty (50) per cent. of any piece of the door. Workmanship must be good, though slight defects therein may be allowed where the quality averages fair. Shaky lumber shall not be admitted. No part of a No. 2 door, except the top rail, short muntins and short panels shall be free from some defect.

No. 3 Doors—No. 3 doors may contain double the amount of defects that are allowed in No. 2, and the knots and other defects may be coarser. Worm eaten lumber may be admitted if showing on one side only. Workmanship may be defective, but not enoughs o to destroy the strength of the door.

WINDOWS.

Check Rail Windows may contain not to exceed two (2) knots in each piece of the sash, said knots not to exceed three-eighths (⅜) inch in diameter. White sap and a small amount of blue sap may be admitted. Workmanship must be good.

Plain Rail Sash may contain blue sap and (small) knots. Shaky lumber not admissible.

BLINDS.

Outside Blinds must be made of clear lumber, except that small pin knots in the stiles and rails, and white sap may be admitted. Shaky lumber is not admitted. Workmanship must be good.

Fig. A.
For Window Frames, Circle inside and outside.

Fig. B.
For Window Frames, Circle outside, Square inside.

PLEASE BE EXPLICIT IN GIVING YOUR ORDERS.

Fig. C.
For Window Frames, Segment inside and outside.

Fig. D.
For Window Frames, Segment outside, Square inside.

Always give the *radius* of the *segment*, otherwise we shall make our regular radius, which is the width of the window.

Windows made in the above forms can be filled with any number of lights required.

PLEASE OBSERVE THE FOLLOWING DIRECTIONS IN GIVING YOUR ORDERS.

DOORS—

Give size, quality (1st or 2nd), thickness and number of panels

MOULDED DOORS—

State whether to be moulded one or two sides, and raised or flush moulding.

SASH—

Give size of glass, number of lights in window, number of lights in width, thickness, plain or check rail, and whether glazed or unglazed.

SEGMENT HEAD SASH—

If frames are made, give radius to inside edge or face casing of window frame, or rise of casing.

ELLIPTIC HEAD SASH—

If frames are made, send a pattern of head casing.

Elliptic, Segment and Circle Sash—

Be sure to state whether the windows are finished square inside or the same shape inside as outside.

OUTSIDE BLINDS—

You are requested to describe the window for which they are intended, following directions given in ordering Sash, and if frames are made, give exact size of opening.

FRAMES—

Give width of jambs for frame buildings and thickness of wall for brick buildings. For door frames state if out or inside frame. For window frames state if plain or check rail, with or without pulleys.

INSIDE BLINDS—

In all cases the order should give the exact size of opening; size and number of lights in window; distance from top of window to center of meeting rail of sash, or to where the blinds are to be cut; number of folds, and whether to be all slats or outside folds paneled; if blinds fold in pockets give width of outside fold.

Do not take it for granted that we know what you want unless you are explicit.

It is economy to conform to regular sizes and styles as much as possible.

When ordering style of work different from what is embraced in list, it is often necessary to give sections, sometimes elevations.

Use terms when practicable, as given in this book.

In giving sizes of Sash, Doors or Blinds, name width first.

A Window indicates two pieces.

A Sash indicates one piece.

A Blind indicates one piece.

A pair of Blinds indicates two pieces.

A set of Sash or Blinds indicates more than two pieces and order should be accompanied with elevation.

HICKS' TELEGRAPH CIPHER

FOR THE

DOOR, SASH AND BLIND TRADE.

The object of this cipher is to enable parties to condense a long telegraph message into a short one by using certain simple words to represent long descriptive terms or long names referring to sizes and styles of manufactured goods. Regular words are used and variations are produced in a regular manner by the use of legitimate endings and prefixes, so as to avoid confusion. Proper nouns are not used except in a legitimate manner when they represent proper nouns, such as the names of railroads.

HOW TO TRANSLATE A CIPHER MESSAGE.

The following alphabetical key will enable everyone to readily translate a message written in cipher by referring to the page indicated. To ensure safety, it is better to look up each word and write out its meaning, so that when translated it lies before you written out in full. Be careful to get each word as it is given:

A—{ Figures.
Plain Rail Sash, Twelve Light.
B—{ Railroads.
Plain Rail Sash, Eight Light.
C—{ Phrases.
Check Rail Sash, Twelve Light.
D—{ De-Prefix.
Check Rail Sash, Eight Light.
E—Check Rail Sash, Four Light.
F—Check Rail Sash, Four Light, Continued.
G—Check Rail Sash, Two Light.
H—Check Rail Sash, Two Light, Continued.
I—Check Rail Sash, Two Light, Continued.
J—Transom Sash.
K—Transom Sash for Double Doors.
L—Pantry Check Rail, Hot Bed and Barn Sash.
M—Cellar Sash.
N—O. G. Four Panel Doors.
O—O. G. Four Panel Doors.
P—O. G. Sash Doors, Sash Doors, Screen and Chamfered Doors.
P G Five Panel Doors.
Q—O G Five Panel Doors.
R—{ Four Panel Moulded Doors.
Re-Prefix.
S—Double Front Doors, Store Doors, Outside Blinds.
U—Un-Prefix.

TELEGRAPH CIPHER.

RAILROADS AND EXPRESS.

Send immediately by	C. & N. W. R. R.	Barton
" " "	C., M. & St. P. R. R.	Bailey
" " "	Wisconsin Central R. R.	Balmoral
" " "	Mil., L. S. & W. R. R.	Baxter
" " "	Mil. & Northern R. R	Bayard
" " "	Union Pacific R. R.	Bingham
" " "	Northern Pacific R. R.	Bacon
" " "	L. & N. R. R.	Belmont
" " "	Illinois Central	Barnum
" " "	L. S. & M. S. R. R.	Barker
" " "	P., Ft. W. & C. R. R.	Bemis
" " "	Kankakee Line	Buffon
" " "	L., N. A. & C. R. R.	Bartlett
" " "	B. & O. R. R.	Bagley
" " "	C., R. I. & P. R. R	Barkman
" " "	C. & E. Ill. R. R.	Brighton
" " "	Chi. & G. T. R. R.	Buckstaff
" " "	P. C. & St. L. R. R	Blakely
" " "	C., B. & Q. R. R	Bunker
" " "	Southern Despatch Line	Bliss
" " "	Michigan Central R. R.	Baker
" " "	C., St. P. M. & O. R. R.	Bolton
" " "	B. & M. R. R	Brooks
" " "	Chicago & Alton R. R.	Butler
" " "	Kansas Pacific R. R	Byford
" " "	Atchison, Topeka & S. F. R. R.	Bly
" " "	Mo. Pacific R. R.	Belle
" " "	Nickel Plate Line	Bellingham
" " "	Erie R. R.	Brookford
" " "	N. Y. Central R. R.	Brockton
" " "	Texas Pacific R. R.	Bass
" " "	Southern Pacific R. R	Bickford
" " "	Morgan R. R.	Bashford
" " "	Express	Blackey
" " "	Way Freight	Briggs
" " "	Merchants' Despatch	Buffington
" " "	Pan-Handle Line	Bradford
" " "	Hoosac Tunnel Line	Bradley
" " "	Georgia Central	Barkis
" " "	Iowa Central	Belford
" " "	Vandalia Line	Belgravia
" " "	Toledo, Wabash & Western	Battis
" " "	M. K. & T. R. R.	Brightwell
" " "	B. C. R. & N.	Berlin

TELEGRAPH CIPHER.

FIGURES.

From 1 to 9 inclusive, spell out the figure thus, *nine.*

10 awl
12 adze
15 ant
20 ass
25 ape
30 aid
35 art
40 alb
45 ache
50 ale
55 ail
60 alms
65 arch
70 arm
75 ash
80 auk
85 awe
90 ax
95 aye
100 abbot
125 abbey
150 abbess
175 abode
200 abscess.
225 abyss
250 absence
275 accent
300 acme
325 acid
350 access
375 acorn
400 action
425 actor
450 actress.
475 acre
500 anchor
525 adage
550 advance
575 adder
600 address
625 adverb
650 adjunct
675 adult
700 advent
725 advice
750 aeric
775 affair
800 afflux
825 affront
850 affray
875 ague
900 agate
925 agent
950 ailment
975 airpump
1,000 airhole
1,100 airgun
1,200 alto
1,300 alarm
1,400 alder
1,500 album
1,600 alley
1,700 almond
1,800 aloe
1,900 alpha
2,000 altar
2,200 alum
2,300 alewife
2,400 almshouse
2,500 amber
2,600 amen
2,700 amuck
2,800 ancient
2,900 angel
3,000 annals
3,100 anise
3,200 ankle
3,300 answer
3,400 antic
3,500 antique
3,600 antler
3,700 anvil
3,800 anxiety
3,900 anthem
4,000 angler
4,100 appeal
4,200 apex
4,300 applause
4,400 apples
4,500 approach
4,600 apron
4,700 aptness
4,800 anger
4,900 arbor
5,000 archness
5,500 arrears
6,000 argus
6,500 armful
7,000 armchair
7,500 armhole
8,000 armor
8,500 armlet
9,000 army
9,500 arcade
10,000 archduke
15,000 archer
20,000 archive
25,000 arctic
30,000 ardor
35,000 argent
40,000 arpent
45,000 arrest
50,000 arrow
55,000 arson
60,000 artist
65,000 aspen
70,000 ascent
75.000 ashes
80,000 ashlar
85,000 aspect
90,000 asphalt
95,000 assault
100,000 assay
200,000 assets
300,000 assize
400,000 aster
500,000 asthma
600,000 athlete
700,000 atlas
800,000 atom
900,000 attic
1,000,000 attire
2,000,000 auction
3,000,000 author
4,000,000 autumn
5,000,000 augur
6,000,000 aurist
7,000,000 awning
8,000,000 axis
9,000,000 axle
10,000,000 aye-aye
20,000,000 azure

TELEGRAPH CIPHER.

PHRASES.

Others have quoted low prices and I must meet the figures..Cabal
Give price on cars at Oshkosh............................Cabbage
Give price in car load lots at Oshkosh........................Cable
Give price on cars at Chicago..............................Cadence
Give price in car load lots at Chicago.........................Cage
Give price delivered here....................................Cactus
Give price in car load lots here..............................Calico
Give price on cars at Mississippi River ....Caligraph
Give price in car load lots at Mississippi River...............Cake
Give price on cars at Missouri River.........................Caitiff
Give price in car load lots at Missouri River..................Calf
Fill ends of car with shingles.............................Calico
Fill ends of cars with lath or pickets.....................Calendar
Order is not good unless filled immediately....................Calk
Order is not good unless filled in ten days.................Cavalry
Prices we give hold good for ten days only ....Calling
Prices we give hold good for this day only.....................Call
I expect to get home Saturday if nothing happens......Calabash
I am out of money; send me about $100 immediately to—.Caloric
Arrived here to-day and find everything satisfactory....Calumny
Arrived here to-day and find everything in bad shape........Calm
Have you in stock and can you ship at once...............Cambric
I think—is hard up and liable to fail. What shall I do?...Camel
Collect if you possibly can, and if you cannot, put it in the hand of a lawyer and enforce collection at once .....Camphor
Come home at once...Canary
It is impossible to fill your order as we are entirely out, and it or they cannot be purchased here................Camphene
Call on————make low prices, and sell him if you possibly can..Cameo
Prices are going up and we are not taking orders except for immediate delivery ....Canal
I will get there Saturday night and spend Sunday there, so you had better send my mail for the next few days to————..Cancer
When will you ship?..Cannon
We have not heard from you in response to our proposition and would like a reply......................................Candy
Answer immediately by telegraph...........................Candle
Answer immediately by mail...............................Candidate
Please remit amount due at once...........................Candor
The account is past due and we will be greatly accommodated by an early settlement..............................Cannibal
If you don't settle at once we shall enforce collection. Please let us hear from you...............................Canoe
We have shut down for repairs and cannot complete your order for a week or two............................Cannonade
We will ship all we have in stock at once.................Canopy
I am sick and will remain here until I hear from you. Notify my family...Canticle
Have written you in full by this mail.......................Canvas

HOW TO WRITE A CIPHER MESSAGE.

Write out the message in ordinary language just as you want it. Then translate into cipher by using whatever is necessary on the pages headed "Phrases," "Railroads" and "Figures," and refer to the different sizes and qualities of goods as they appear in their proper place in the Price List.

EXAMPLE.

PLATTE VALLEY, NEB., March 1, 1888.

Carlton Foster & Co., Chicago, Illinois:

Send us immediately by Union Pacific Railroad, seventy-five Twelve Light Windows, Plain Rail Sash, 8 x 12, one and one-eighth thick, glazed single strength, forty-five Four Panel Moulded Doors, two eight, six eight, one and three-eighths thick, raised moulding on two sides. Fill ends of car with shingles. Have written you in full by this mail. Answer immediately by mail.

JOHN SAWLOG.

65 words.

By using the cipher this message of 65 words can be condensed into a message of 8 words, as follows:

PLATTE VALLEY, NEB., March 1, 1888.

Carlton Foster & Co., Chicago, Illinois:

Bingham ash abashed, ache repress capstan canvas candidate.

8 words. JOHN SAWLOG.

EXAMPLE.

CHICAGO, ILL., Jan. 25, 1888.

Messrs. Mortise & Tenon, Boulder, Col.:

Camphene canal candy candle calling canvas.

6 paid. CARLTON FOSTER & CO.

Translated the above message reads as follows:

CHICAGO, ILL., Jan. 25, 1888.

Messrs. Mortise & Tenon, Bowlder, Col.:

It is impossible to fill your order, as we are entirely out, and it cannot be purchased here. Prices are going up, and we are not taking orders except for immediate delivery. We have not heard from you in response to our proposition, and would like a reply. Answer immediately by telegraph. Prices we give hold good for ten days only. Have written you in full by this mail.

70 words. CARLTON FOSTER & CO.

Cipher is for Open Sash.
If Glazed S. S. add *ed.*
1⅜ add *ly.* 1¾ add *ism.*
Circle, prefix *dis.*
Segment, prefix *re.*

PLAIN RAIL SASH.

Twelve-Lighted Windows.

Cipher	Size of Glass.	Thickness.	Price per Window Open.	Price per Window Glazed.	Size of Window.
	Inches.	Inches.	$ cts.	$ cts.	Ft. In. Ft. In.
Abduct	7× 9	1 1-8	.46	1.35	2 1 × 3 4½
Abandon	8×10	"	.55	1.60	2 4 × 3 9¼
Abash	8×12	"	.64	2.00	2 4 × 4 6
Abolish	8×14	"	.70	2.35	2 4 × 5 2
Abound	8×16	"	.77	2.70	2 4 × 5 10
Abscond	9×12	"	.72	2.20	2 7 × 4 6
Abstain	9×13	"	.75	2.35	2 7 × 4 10
Absorb	9×14	"	.75	2.45	2 7 × 5 2
Absent	9×15	"	.79	2.70	2 7 × 5 6
Abstract	9×16	"	.81	2.80	2 7 × 5 10
Accent	9×18	"	.86	3.60	2 7 × 6 6
Accomplish	10×12	"	.74	2.40	2 10 × 4 6
Account	10×14	"	.77	2.70	2 10 × 5 2
Accord	10×15	"	.81	2.95	2 10 × 5 6
Accroach	10×16	"	.81	3.30	2 10 × 5 10
Accredit	10×18	"	.86	3.65	2 10 × 6 6
Accustom	10×20	"	.98	4.25	2 10 × 7 2
Acquaint	12×14	"	.97	3.80	3 4 × 5 2
Act	12×16	"	1.03	4.25	3 4 × 5 10
Affect	12×18	"	1.10	4.65	3 4 × 6 6

1 3-8 Plain Rail Windows same price as 1 3-8 Check Windows.

PLAIN RAIL SASH.

Eight-Lighted Windows.

Size of Glass.	Thickness.	Price per Window Open.	Price per Window Glazed.	Size of Window.	*Cipher* is for Open Sash. If Glazed S. S. add *ed*. Glazed D. S. add *edly*. 1¾ prefix *dis*.
Inches.	Inches.	$ cts.	$ cts.	Ft. In. Ft. In.	
8×10	1 1-8	.52	1.25	1 8⅛×3 9¼	Back
8×12	"	.58	1.45	1 8⅛×4 6	Badger
8×14	"	.65	1.75	1 8⅛×5 2	Bail
8×16	"	.69	1.95	1 8⅛×5 10	Balk
9×12	"	.65	1.60	1 10⅛×4 6	Band
9×14	"	.69	1.85	1 10⅛×5 2	Banish
9×16	"	.74	2.15	1 10⅛×5 10	Bang
10×12	"	.67	1.80	2 0⅛×4 6	Bank
10×14	"	.70	1.95	2 0⅛×5 2	Banter
10×16	"	.75	2.40	2 0⅛×5 10	Barb
10×18	"	.84	2.80	2 0⅛×6 6	Banquet
12×14	"	.80	2.55	2 4⅛×5 2	Bankrupt
12×16	"	.88	2.85	2 4⅛×5 10	Bargain
12×18	"	.94	3.15	2 4⅛×6 6	Bark

1 3-8 Plain Rail Windows same price as 1 3-8 Check Windows.

CHECK RAIL SASH.

Twelve-Lighted Windows.

Cipher is for Open Sash.
For Glazed S. S. add *ed.*
Glazed D. S. add *edly.*
1¾ inch, prefix *re.*
Segment, prefix *un.*
Circle, prefix *dis.*

Cipher	Size of Glass.	Thickness.	Price per Window Open.	Price per Window Glazed.	Size of Window.
	Inches.	Inches.	$ cts.	$ cts.	Ft. In Ft In.
Connect	8×10	1 3-8	.73	1.90	2 4⅛ × 3 10
Commend	8×12	"	.88	2.30	2 4⅛ × 4 6
Compress	8×14	"	1.03	2.70	2 4⅛ × 5 2
Concern	8×16	"	1.15	3.05	2 4⅛ × 5 10
Conduct	9×12	"	1.01	2.55	2 7⅛ × 4 6
Conform	9×13	"	1.10	2.70	2 7⅛ × 4 10
Conflict	9×14	"	1.13	2.90	2 7⅛ × 5 2
Confront	9×15	"	1.18	3.05	2 7⅛ × 5 6
Confess	9×16	"	1.22	3.20	2 7⅛ × 5 10
Conjoin	9×18	"	1.35	4.05	2 7⅛ × 6 6
Conscript	10×12	"	1.12	2.80	2 10⅛ × 4 6
Consign	10×14	"	1.21	3.15	2 10⅛ × 5 2
Consort	10×15	"	1.26	3.40	2 10⅛ × 5 6
Constrict	10×16	"	1.27	3.80	2 10⅛ × 5 10
Constrain	10×18	"	1.35	4.15	2 10⅛ × 6 6
Consult	10×20	"	1.67	5.10	2 10⅛ × 7 2
Content	10×22	"	1.90	5.60	2 10⅛ × 7 10
Condemn	10×24	"	2.20	6.25	2 10⅛ × 8 6
Contort	12×14	"	1.57	4 40	3 4⅛ × 5 2
Contract	12×16	"	1.64	4.85	3 4⅛ × 5 10
Convert	12×18	"	1.76	5.35	3 4⅛ × 6 6
Convict	12×20	"	1.94	6.00	3 4⅛ × 7 2
Convey	12×22	"	2.15	6.65	3 4⅛ × 7 10
Correct	12×24	"	2.42	7.30	3 4⅛ × 8 6

For price of 1¾ inch Windows, add 75 cents each to above list.
All Check Windows plowed and bored.

CHECK RAIL SASH.

Eight-Lighted Windows.

Cipher is for Open Sash. For Glazed S. S. add *ed.* Glazed D. S. add *edly.* Segment, prefix *un.* Circle, prefix *re.*

Size of Glass.	Thickness.	Price per Window Open.	Price per Window Glazed.	Size of Window.	Cipher
Inches.	Inches.	$ cts.	$ cts.	Ft. In. Ft. In	
9×12	1 3-8	.69	1.70	1 11 × 4 6	Delight
9×14	"	.82	2.00	1 11 × 5 2	Design
9×16	"	.88	2.30	1 11 × 5 10	Deflower
9×18	"	.99	2.80	1 11 × 6 6	Demand
10×12	"	.70	1.85	2 1 × 4 6	Deny
10×14	"	.83	2.15	2 1 × 5 2	Disdain
10×16	"	.92	2.60	2 1 × 5 10	Demean
10×18	"	1.05	3.05	2 1 × 6 6	Deliver
10×20	"	1.18	3.45	2 1 × 7 2	Dismiss
12×14	"	.85	2.60	2 5 × 5 2	Depress
12×16	"	.95	2.95	2 5 × 5 10	Descry
12×18	"	1.07	3.30	2 5 × 6 6	Despoil
12×20	"	1.26	3.85	2 5 × 7 2	Devour
14×16	"	1.21	3.65	2 9 × 5 10	Discard
14×18	"	1.23	3.80	2 9 × 6 6	Disclaim
14×20	"	1.38	4.20	2 9 × 7 2	Disburden
14×22	"	1.57	4.65	2 9 × 7 10	Decoy
14×24	"	1.82	5.20	2 9 × 8 6	Depart

Continued on next page.

Segment Head 1⅜ thick, add 60c; 1¾ thick, add 80c.
Half Circle Head, 1⅜ thick, add $1.20; 1¾ thick, add $1.50.
Sizes not given above, extra price.
All Check Windows plowed and bored.

FIFTEEN AND EIGHTEEN LIGHT WINDOWS.

For 15 Light Windows, add to price of 12 Light, one-third.
For 18 Light Windows, add to price of 12 Light, two-thirds

CHECK RAIL SASH.

Eight-Lighted Windows—Continued.

Cipher is for Open Sash. For Glazed S. S. add *ed.* Glazed D. S. add *edly.* Segment, prefix *un.* Circle, prefix *re.*	Size of Glass.	Thickness	Price per Window Open.	Price per Window Glazed.	Size of Window.
	Inches.	Inches	$ cts.	$ cts.	Ft. In. Ft. In.
Detect	10×16	1 3-4	1.63	3.45	2 1 × 5 10
Detail	10×18	"	1.70	3.75	2 1 × 6 6
Detain	10×20	"	1.82	4.10	2 1 × 7 2
Desist	12×14	"	1.60	3.50	2 5 × 5 2
Discant	12×16	"	1.70	3.85	2 5 × 5 10
Depict	12×18	"	1.82	4.20	2 5 × 6 6
Default	12×20	"	1.95	4.65	2 5 × 7 2
Decoct	12×22	"	2.15	5.15	2 5 × 7 10
Discount	12×24	"	2.35	5.60	2 5 × 8 6
Declaim	14×16	"	1.90	4.45	2 9 × 5 10
Deduct	14×18	"	1.98	4.80	2 9 × 6 6
Demolish	14×20	"	2.13	5.25	2 9 × 7 2
Descend	14×22	"	2.32	5.70	2 9 × 7 10
Divest	14×24	"	2.57	6.25	2 9 × 8 6

Segment Head, 1⅜ thick, add 60c; 1¾ thick, add 80c.

Half Circle Head, 1⅜ thick, add $1.20; 1¾ thick, add $1.50.

All check windows plowed and bored.

Sizes not given above, extra price.

N. B.—We make **Queen Anne Sash** of all kinds, designs of which will be furnished upon application.

When ordering **Bent Sash** for tower window be sure and send plan for same.

CHECK RAIL SASH.

Four Lighted Windows.

Cipher is for Open Sash.
For Glazed S. S. add *ed.*
Glazed D. S. add *edly.*
Segment, prefix *un.*
Circle, prefix *re.*

Size of Glass.	Thickness.	Price per Window Open.	Price per Window Glazed.	Size of Window.	
Inches.	Inches.	$ cts.	$ cts.	Ft. In. Ft. In.	
10×20	1 3-8	.58	1.60	2 1 × 3 10	Evict
10×22	"	.65	1.90	2 1 × 4 2	Elect
10×24	"	.69	1.95	2 1 × 4 6	Embark
10×26	"	.76	2.10	2 1 × 4 10	Employ
10×28	"	.81	2.25	2 1 × 5 2	Engross
10×30	"	.84	2.55	2 1 × 5 6	Enlist
10×32	"	.90	2.90	2 1 × 5 10	Emboss
10×34	"	1.00	3.30	2 1 × 6 2	Enjoin
10×36	"	1.05	3.45	2 1 × 6 6	Entrust
12×20	"	.60	1.85	2 5 × 3 10	Entrench
12×22	"	.65	2.00	2 5 × 4 2	Entreat
12×24	"	.72	2.20	2 5 × 4 6	Escort
12×26	"	.80	2.40	2 5 × 4 10	Esteem
12×28	"	.85	2.70	2 5 × 5 2	Exalt
12×30	"	.88	3.20	2 5 × 5 6	Exceed
12×32	"	.95	3.35	2 5 × 5 10	Exhibit
12×34	"	1.00	3.60	2 5 × 6 2	Expand
12×36	"	1.07	3.85	2 5 × 6 6	Exist
12×38	"	1.13	4.05	2 5 × 6 10	Express
12×40	"	1.20	4.55	2 5 × 7 2	Extend

Continued on next page.

Cipher is for Open Sash.
For Glazed S. S. add *ed*.
Glazed D. S. add *edly*.
Segment, prefix *un*.
Circle, prefix *re*.
1¾ inch, prefix *dis*.

CHECK RAIL SASH.

Four-Lighted Windows.—Continued.

	Size of Glass.	Thickness.	Price per Window Open.	Price per Window Glazed.	Glazed Double Str'n'h	Size Window.
	Inches.	Inches	$ cts.	$ cts.	$ cts.	Ft. In. Ft. In.
Extort	14 × 24	1 3-8	1.08	2.85	....	2 9 × 4 6
Exult	14 × 26	"	1.10	3.05	....	2 9 × 4 10
Extract	14 × 28	"	1.10	3.55	....	2 9 × 5 2
Enter	14 × 30	"	1.13	3.90	....	2 9 × 5 6
Fasten	14 × 32	"	1.15	4.10	5.35	2 9 × 5 10
Fulfill	14 × 34	"	1.20	4.35	5.70	2 9 × 6 2
Fill	14 × 36	"	1.23	4.50	6.05	2 9 × 6 6
Flash	14 × 38	"	1.36	5.10	6.55	2 9 × 6 10
Frill	14 × 40	"	1.38	5.30	6.95	2 9 × 7 2
Flutter	14 × 42	"	1.58	6.35	8.15	2 9 × 7 6
Fold	14 × 44	"	1.66	6.40	8.25	2 9 × 7 10
Follow	14 × 46	"	1.87	7.30	9.05	2 9 × 8 2
Forfeit	14 × 48	"	1.92	7.50	9.70	2 9 × 8 6
Fawn	15 × 24	"	1.30	3.30	3.95	2 11 × 4 6
Fallow	15 × 26	"	1.33	4.05	4.90	2 11 × 4 10
Falter	15 × 28	"	1.35	4.25	5.35	2 11 × 5 2
Famish	15 × 30	"	1.38	4.60	5.60	2 11 × 5 6
Farm	15 × 32	"	1.40	4.70	5.90	2 11 × 5 10
Farrow	15 × 34	"	1.45	5.15	6.30	2 11 × 6 2
Fashion	15 × 36	"	1.50	5.75	7.10	2 11 × 6 6
Feast	15 × 38	"	1.55	5.80	7.15	2 11 × 6 10
Father	15 × 40	"	1.65	6.65	8.20	2 11 × 7 2
Fathom	15 × 42	"	1.75	7.20	8.90	2 11 × 7 6
Fatten	15 × 44	"	1.85	7.30	9.00	2 11 × 7 10
Fault	15 × 46	"	2.00	8.40	10.60	2 11 × 8 2
Favor	15 × 48	"	2.10	8.50	10.70	2 11 × 8 6

Continued on next page.

Cipher is for Open Sash.
For Glazed S. S. add *ed.*
Glazed D. S. add *edly.*
Segment, prefix *un.*
Circle, prefix *re.*
1¾ inch prefix, *dis.*

CHECK RAIL SASH.

Four-Lighted Windows.—Continued.

Size of Glass.	Thickness.	Price per Window Open.	Price per Window. Glazed.	Glazed Double Str'n'h	Size Window.	
Inches.	Inches	$ cts.	$ cts.	$ cts.	Ft In. Ft. In.	
12 × 30	1 3-4	1.63	4.05	5.00	2 5 × 5 6	Form
12 × 32	"	1.70	4.25	5.25	2 5 × 5 10	Filter
12 × 34	"	1.75	4.50	5.50	2 5 × 6 2	Flatten
12 × 36	"	1.82	4.70	5.80	2 5 × 6 6	Ferment
12 × 38	"	1.88	4.95	6.10	2 5 × 6 10	Forestall
12 × 40	"	1.95	5.45	6.80	2 5 × 7 2	Fowl
14 × 30	"	1.88	4.75	5.85	2 9 × 5 6	Foreshadow
14 × 32	"	1.90	4.95	6.10	2 9 × 5 10	Flavor
14 × 34	"	1.95	5.25	6.45	2 9 × 6 2	Float
14 × 36	"	1.98	5.50	6.80	2 9 × 6 6	Flicker
14 × 38	"	2.05	5.80	7.20	2 9 × 6 10	Finish,
14 × 40	"	2.13	6.25	7.70	2 9 × 7 2	Festoon
14 × 42	"	2.23	7.00	8.80	2 9 × 7 6	Feign
14 × 44	"	2.32	7.10	8.90	2 9 × 7 10	Fear
14 × 46	"	2.45	7.60	9.60	2 9 × 8 2	Faint
14 × 48	"	2.57	8.15	10.40	2 9 × 8 6	Fail

Windows with glass 11 inches wide same price as 12 inch,

Windows with glass 13 inch and 13⅛ inch same price as 14 inch.

For Windows 1¾ inch, where not listed, add 75c to list of 1⅜ inch.

Segment Head, 1⅜ thick, add 60c; 1¾ thick, add 80c. Half Circle Head, 1⅜ thick, add $1.20; 1¾ thick, add $1.50. Sizes not given, extra price. All four-light windows plowed and bored.

To arrive at price of 1⅜ Four-Light Windows, glazed double strength, deduct from 1¾ list the difference between 1⅜ and 1¾ Open Sash.

Cipher is for Open Sash.
For Glazed S. S. add *ed.*
Glazed D. S. add *edly.*
Segment, prefix *un.*
Circle, prefix *re.*
1¾in. not given, prefix *dis*

CHECK RAIL SASH.

Two-Lighted Windows.

	Size of Glass.	Thick-ness.	Price wind'w open.	Glazed Single Strength.	Glazed Double Strength.	Size Window.
	Inches.	Inches	$ cts.	$ cts.	$ cts.	Ft. In. Ft. In.
Govern	16 × 24	1 3-8	.70	1.75	2.10	1 8⅛×4 6
Grain	16 × 26	"	.78	2.20	2.80	1 8⅛×4 10
Ground	16 × 28	"	.78	2.30	2.90	1 8⅛×5 2
Gird	16 × 30	"	.80	2.35	3.05	1 8⅛×5 6
Glisten	16 × 32	"	.85	2.50	3.25	1 8⅛×5 10
Guard	16 × 34	"	.97	2.90	3.60	1 8⅛×6 2
Grant	16 × 36	"	.97	3.05	3.75	1 8⅛×6 6
Greet	18 × 24	"	.70	2.25	2.70	1 10⅛×4 6
Glitter	18 × 26	"	.81	2.40	2.90	1 10⅛×4 10
Gild	18 × 28	"	.82	2.55	3.25	1 10⅛×5 2
Gather	18 × 30	"	.85	2.60	3.30	1 10⅛×5 6
Gain	18 × 32	"	.92	2.80	3.55	1 10⅛×5 10
Gallop	18 × 34	"	.98	3.20	4.00	1 10⅛×6 2
Garnish	18 × 36	"	1.02	3.40	4.30	1 10⅛×6 6
Garrison	18 × 38	"	1.10	3.80	4.70	1 10⅛×6 10
Gall	18 × 40	"	1.18	4.15	5.15	1 10⅛×7 2
Gossip	20 × 24	"	.70	2.25	2.85	2 0⅛×4 6
Glint	20 × 26	"	.81	2.45	3.10	2 0⅛×4 10
Guess	20 × 28	"	.82	2.60	3.30	2 0⅛×5 2
Gleam	20 × 30	"	.85	2.75	3.50	2 0⅛×5 6
Gash	20 × 32	"	.92	3.20	4.05	2 0⅛×5 10
Hack	20 × 34	"	1.00	3.30	4.15	2 0⅛×6 2
Hail	20 × 36	"	1.07	3.80	4.85	2 0⅛×6 6
Hammer	20 × 38	"	1.19	3.90	4.95	2 0⅛×6 10
Hand	20 × 40	"	1.20	4.25	5.40	2 0⅛×7 2

Continued on next page.

CHECK RAIL SASH.

Two-Lighted Windows—Continued.

Size of Glass.	Thickness.	Price Window Open.	Glazed Single Strength	Glazed Double Strength	Size Window.	*Cipher* is for Open Sash. For Glazed S. S. add *ed.* Glazed D. S. add *edly.* Segment, prefix *un.* Circle, prefix *re.* 1¾ in. prefix *dis*
Inches.	Inches.	$ cts.	$ cts.	$ cts.	Ft. In. Ft. In.	
22 × 24	1 3-8	.76	2.40	3.05	2 2⅛×4 6	Hang
22 × 26	"	.84	2.60	3.30	2 2⅛×4 10	Harden
22 × 28	"	.85	2.75	3.50	2 2⅛×5 2	Harbor
22 × 30	"	.88	3.20	4.00	2 2⅛×5 6	Harness
22 × 32	"	.95	3.45	4.35	2 2⅛×5 10	Harvest
22 × 34	"	1.00	3.70	4.75	2 2⅛×6 2	Hatch
22 × 36	"	1.07	4.05	5.20	2 2⅛×6 6	Help
22 × 38	"	1.19	4.20	5.35	2 2⅛×6 10	Harpoon
22 × 40	"	1.20	4.85	6.30	2 2⅛×7 2	Herd
24 × 28	"	.85	3.15	3.95	2 4⅛×5 2	Hill
24 × 30	"	.88	3.40	4.30	2 4⅛×5 6	Hinder
24 × 32	"	.95	3.65	4.70	2 4⅛×5 10	Hiss
24 × 34	"	1.00	4.00	5.15	2 4⅛×6 2	Hoard
24 × 36	"	1.07	4.05	5.20	2 4⅛×6 6	Honor
24 × 38	"	1.19	4.80	6.25	2 4⅛×6 10	Hurl
24 × 40	"	1.20	4.85	6.30	2 4⅛×7 2	Husband
24 × 42	"	1.40	5.55	7.25	2 4⅛×7 6	Hunt
24 × 44	"	1.49	5.65	7.30	2 4⅛×7 10	Imbitter
26 × 30	"	1.19	4.20	5.35	2 6⅛×5 6	Impact
26 × 32	"	1.21	4.20	5.40	2 6⅛×5 10	Impair
26 × 34	"	1.26	4.65	5.95	2 6⅛×6 2	Impark
26 × 36	"	1.30	4.95	6.40	2 6⅛×6 6	Induct
26 × 38	"	1.38	5.55	7.20	2 6⅛×6 10	Import
26 × 40	"	1.46	5.60	7.30	2 6⅛×7 2	Impeach
26 × 42	"	1.52	5.70	7.35	2 6⅛×7 6	Imprint
26 × 44	"	1.60	6.45	8.40	2 6⅛×7 10	Implant
26 × 46	"	1.69	7.15	8.85	2 6⅛×8 2	Impugn
26 × 48	"	1.76	7.20	8.90	2 6⅛×8 6	Indent

Continued on next page.

Cipher is for Open Sash.
For Glazed S. S. add *ed.*
Glazed D. S. add *edly.*
Segment, prefix *un.*
Circle, prefix *re.*
1¾ in. prefix *dis.*

CHECK RAIL SASH.

Two-Lighted Windows.—Continued.

Cipher	Size of Glass.	Thick-ness.	Price Window Open.	Glazed Single Strength	Glazed Double Strength	Size Window.
	Inches.	Inches.	$ cts.	$ cts.	$ cts.	Ft. In. Ft. In.
Index	28 × 30	1 3-8	1.20	4.25	5.40	2 8⅛×5 6
Infest	28 × 32	"	1.21	4.60	5.90	2 8⅛×5 10
Infix	28 × 34	"	1.26	4.90	6.35	2 8⅛×6 2
Inflict	28 × 36	"	1.30	5.45	7.10	2 8⅛×6 6
Inherit	28 × 38	"	1.38	5.55	7.20	2 8⅛×6 10
Inhabit	28 × 40	"	1.46	5.60	7.30	2 8⅛×7 2
Ingrain	28 × 42	"	1.52	6.35	8.30	2 8⅛×7 6
Inject	28 × 44	"	1.60	7.05	8.75	2 8⅛×7 10
Instill	28 × 46	"	1.69	7.15	8.85	2 8⅛×8 2
Insult	28 × 48	"	1.76	8.25	10.35	2 8⅛×8 6
Intend	30 × 32	"	1.45	5.10	6.55	2 10⅛×5 10
Intercept	30 × 34	"	1.55	5.70	7.35	2 10⅛×6 2
Interchain	30 × 36	"	1.63	5.80	7.45	2 10⅛×6 6
Interdict	30 × 38	"	1.72	5.90	7.50	2 10⅛×6 10
Interest	30 × 40	"	1.84	6.70	8.60	2 10⅛×7 2
Interject	30 × 42	"	1.92	7.35	9.10	2 10⅛×7 6
Interlink	30 × 44	"	2.03	7.45	9.20	2 10⅛×7 10
Interlock	30 × 46	"	2.14	8.65	10.70	2 10⅛×8 2
Intermix	30 × 48	"	2.25	8.75	10.80	2 10⅛×8 6
Interplead	30 × 50	"	2.38	8.90	10.95	2 10⅛×8 10

For windows 1¾ inch, add 75 cents each.
Segment Head, 1⅜ thick, add 60c; 1¾ thick, add 80c.
Half Circle Head, 1⅜ thick, add $1.20; 1¾ thick, add $1.50.
Sizes not given above, extra price.
In ordering Sash, Doors and Blinds always give the width first.
Use terms, when practicable, as given in this book.
All Check Windows plowed and bored.

Cipher is for Open Sash.
For Glazed S. S. add *ed.*
Glazed D. S. add *edly.*
1¾ add *ism*,
Segment, prefix *un.*
Circle, prefix *dis.*

PANTRY CHECK RAIL SASH.

Four-Lighted Windows, One Light Wide.

Size of Glass.	Thickness.	Price per Window Open.	Price per Window Glazed.	Size of Window.	Cipher
Inches.	Inches.	$ cts.	$ cts.	Ft. In. Ft. In.	
9 × 12	1 3-8	.63	1.40	1 1⅛ × 4 6	Languish
9 × 14	"	.70	1.65	1 1⅛ × 5 2	Label
9 × 16	"	.75	1.85	1 1⅛ × 5 10	Lack
12 × 14	"	.73	1.85	1 4⅛ × 5 2	Lampoon
12 × 16	"	.80	2.05	1 4⅛ × 5 10	Launch
12 × 18	"	.87	2.20	1 4⅛ × 6 6	Land

Two-Lighted Windows, One Light Wide.

Size of Glass.	Thickness.	Price per Window Open.	Price per Window Glazed.	Size of Window.	Cipher
Inches.	Inches.	$ cts.	$ cts.	Ft. In. Ft. In.	
12 × 24	1 3-8	.60	1.60	1 4⅛ × 4 6	Lash
12 × 26	"	.68	1.80	1 4⅛ × 4 10	Latch
12 × 28	"	.73	1.85	1 4⅛ × 5 2	Last
12 × 30	"	.75	2.20	1 4⅛ × 5 6	Lead
12 × 32	"	.80	2.30	1 4⅛ × 5 10	Learn
12 × 34	"	.85	2.40	1 4⅛ × 6 2	Lift
12 × 36	"	.87	2.50	1 4⅛ × 6 6	Level
14 × 24	"	.70	1.70	1 6⅛ × 4 6	Litter
14 × 26	"	.75	1.85	1 6⅛ × 4 10	Light
14 × 28	"	.75	2.15	1 6⅛ × 5 2	Limit
14 × 30	"	.77	2.30	1 6⅛ × 5 6	Linger
14 × 32	"	.83	2.40	1 6⅛ × 5 10	List
14 × 34	"	.88	2.70	1 6⅛ × 6 2	Lock
14 × 36	"	.90	2.80	1 6⅛ × 6 6	Leap

All Check Windows plowed and bored.

Cypher is for Open Sash.
For Glazed S. S. add *ed*.
Glazed D. B. add *edly*.
1¾ inch prefix *dis*.
1 Light, add *ing*.

TRANSOM SASH.

1 3-8—One and Two Lights.

	Size of Glass.	Price per Sash Open.	Price Two Light Glazed.	Price One Light Glazed.	Size of Sash.
	Inches.	$ cts.	$ cts.	$ cts.	Ft. In. In.
Jerk	13 × 8	.32	.75	.95	2 6 × 12
Jeer	13 × 10	.32	.80	1.00	2 6 × 14
Jest	13 × 12	.32	.90	1.10	2 6 × 16
Join	14 × 8	.40	.90	1.10	2 8 × 12
Jump	14 × 10	.40	.95	1.15	2 8 × 14
Joist	14 × 12	.40	1.10	1.30	2 8 × 16
Junket	15 × 10	.40	.95	1.15	2 10 × 14
Joy	15 × 12	.40	1.15	1.35	2 10 × 16
Joint	15 × 14	.40	1.25	1.45	2 10 × 18
Jilt	16 × 10	.40	1.05	1.25	3 0 × 14
Jewel	16 × 12	.40	1.15	1.35	3 0 × 16
Jaunt	16 × 14	.40	1.25	1.45	3 0 × 18
Jabber	16 × 16	.48	1.45	1.65	3 0 × 20

SEGMENT TRANSOM SASH.

One Light for Double Doors.

	Size of Sash.	Thickness.	Price Open.	Price Glazed.
	Ft. In. In.	Inches.	$ cts.	$ cts.
Keel	4 0 × 18	1 3-4	1.38	4.45
Kick	4 4 × 20	"	1.57	4.35
Kiss	4 6 × 18	"	1.63	4.20
Kill	4 6 × 20	"	1.70	4.55
Knight	4 6 × 24	"	1.75	5.25
Knock	5 0 × 22	"	1.88	5.70
Kneel	5 0 × 24	"	2.00	6.40

For Square Top Transoms deduct 65c from above list.

CELLAR SASH.

Three Lights.

Size of Glass.	Thickness.	Price Open.	Price Single Glazed.	Price Double Glazed.	Size of Sash.	Cipher
Inches.	Inches.	$ cts.	$ cts.	$ cts.	Ft. In. Ft. In.	
7 × 9	1 1-8	.20	.60	.85	2 1 × 1 1	Maul
8 × 10	"	.25	.75	1.05	2 4 × 1 2	Madden
9 × 12	"	.28	.85	1.25	2 7 × 1 4	Maim
9 × 13	"	.28	.90	1.30	2 7 × 1 5	Maintain
9 × 14	"	.32	1.00	1.40	2 7 × 1 6	Malt
9 × 15	"	.32	1.00	1.45	2 7 × 1 7	Mark
9 × 16	"	.38	1.15	1.75	2 7 × 1 8	Marvel
10 × 12	"	.32	.95	1.35	2 10 × 1 4	Master
10 × 14	"	.32	1.00	1.45	2 10 × 1 6	Matter
10 × 15	"	.32	1.00	1.50	2 10 × 1 7	Match
10 × 16	"	.38	1.20	1.85	2 10 × 1 8	Melt
12 × 12	"	.32	1.00	1.45	3 4 × 1 4	Mention
12 × 14	"	.38	1.25	1.90	3 4 × 1 6	Merit
12 × 16	"	.40	1.35	2.10	3 4 × 1 8	Mess
12 × 18	"	.45	1.50	2.30	3 4 × 1 10	Mew

Cipher is for Open Sash. For Glazed S. S. add *ed*. Glazed D. S. add *edly*. Double Rabbetted pre. *un*.

HOT BED SASH.

Made for 6 or 7 inch Glass. Odd Sizes Extra Price.

Size of Sash.	Thickness.	Price each Sash Open.	Price each Sash Glazed.	Cipher
Ft. In. Ft. In.	Inches.	$ cts.	$ cts.	
3 0 × 6 0	1 3-8	2.00	4.25	Lump
3 0 × 6 0	1 3-4	2.50	5.00	Lynch

Cipher is for Open Sash. For Glazed S. S. add *ed*.

STORM SASH.

For price of Storm Sash add to Check Rail Windows 50c each.
Ventilator in bottom add 15 cents net.
Swinging light add 40 cents net.

CELLAR SASH.

Two-Lighted Windows.

Cipher is for Open Sash.
For Glazed S. S. add *ed.*
Glazed D. S. add *edly.*

Cipher	Size of Glass.	Thickness.	Price Open.	Price Single Glazed.	Price Double Glazed.	Size of Sash.
	Inches.	Inches.	$ cts.	$ cts.	$ cts.	Ft. In. In.
Mildew	10×12	1 3-8	.32	.85	....	2 1 × 16
Milk	10×14	"	.32	.90	......	2 1 × 18
Mill	10×16	"	.40	1.05	....	2 1 × 20
Mind	10×18	"	.40	1.10	....	2 1 × 22
Minister	12×12	"	.35	.90	....	2 5 × 16
Mint	12×14	"	.40	1.10	....	2 5 × 18
Mirror	12×16	"	.40	1.15	....	2 5 × 20
Miscall	12×18	"	.45	1.25	....	2 5 × 22
Misdeem	12×20	"	.50	1.35	....	2 5 × 24
Misdirect	14×16	"	.45	1.25	....	2 9 × 20
Misform	14×18	"	.50	1.40	....	2 9 × 22
Misgovern	14×20	"	.55	1.50	....	2 9 × 24
Misinterpret	14×22	"	.60	1.60	....	2 9 × 26
Misinstruct	14×24	"	.65	1.70	....	2 9 × 28

FOUR-LIGHT BARN SASH.

Cipher is for Unglazed.
For Glazed S. S. add *ed.*
Glazed D. S. add *edly.*
1¾ inch, prefix *un.*

Cipher	Size of Glass.	Thickness.	Price per Sash Open.	Price per Sash Glazed.
	Inches.	Inches.	$ cts.	$ cts.
Lunch	8 × 10	1 1-8	.32	.90
Luff	9 × 12	"	.38	1.05
Labor	9 × 14	"	.45	1.30
Lord	9 × 16	"	.45	1.45
Look	10 × 12	"	.43	1.20
Loiter	10 × 14	"	.50	1.35

WIRE SCREEN DOORS.

Covered with best Wire Cloth.
Raised Moulding outside, O G finish inside.

SIZE.					Thickness	Price.	Stained Walnut, add *ing.* Frames, no wire, add *ly.*
Ft.	In.		Ft.	In.	Inches.	$ cts.	
2	6	×	6	6	1 1-8	2.50	Pour
2	8	×	6	8	"	2.65	Preach
2	10	×	6	10	"	2.85	Prank
3	0	×	7	0	"	3.15	Prick

Stained, imitation walnut or painted green.

We make all our stock sizes ⅛ inch each longer and wider than list, to allow for fitting. Hardwood frames made to order.

WINDOW SCREENS MADE TO ORDER.

Pine Frames, complete with wire netting....17c per square foot.

Subject to Discount.

Cipher is for 1st Quality.
For 2d Qual. add *ed* or *d*.
3d Qual. add *edly* or *dly*.

O G FOUR-PANEL DOORS.

Raised Panels Both Sides.

	SIZE.	Thickness.	Price First Quality.	Price 2nd Quality.	Price 3rd Quality.
	Ft. In. Ft. In.	Inches.	$ cts.	$ cts.	$ cts
Narrow	2 0 × 6 0	1 1-8	2.20	2.05	
Near	2 4 × 6 4	"	2.40	2.25	
Need	2 0 × 6 6	"	2.50	2.35	
Neglect	2 6 × 6 6	"	2.50	2.35	2.00
Neigh	2 6 × 6 8	"	2.65	2.50	
Nourish	2 8 × 6 8	"	2.65	2.50	2.15
Number	2 10 × 6 10	"	2.90	2.75	2.30
Nick	3 0 × 7 0	"	3.15	3.00	2.50
Nest	2 0 × 6 0	1 3-8	2.50	2.25	
Neighbor	2 4 × 6 4	"	2.65	2.40	
Nonsuit	2 6 × 6 6	"	2.80	2.55	2.20
Nail	2 6 × 6 8	"	3.00	2.70	
Notch	2 8 × 6 8	"	3.00	2.70	2.30
Negative	2 6 × 6 10	"	3.20	2.90	
Nettle	2 8 × 6 10	"	3.25	2.95	
Needle	2 10 × 6 10	"	3.30	3.00	2.50
Navigate	2 0 × 7 0	"	3.40	3.10	
Name	2 6 × 7 0	"	3.45	3.15	
Neutralize	2 8 × 7 0	"	3.50	3.20	
Nominate	2 10 × 7 0	"	3.55	3.25	
Nose	3 0 × 7 0	"	3.60	3.30	2.80
Notice	2 8 × 7 6	"	4.15	3.85	
Numerate	2 10 × 7 6	"	4.20	3.90	
Nurture	3 0 × 7 6	"	4.25	3.95	
Nuzzle	3 0 × 8 0	"	4.90	4.60	
Nurse	3 0 × 8 6	"	5.80	5.50	
Niggle	3 0 × 9 0	"	6.70	6.40	

Continued on next page.

O G FOUR-PANEL DOORS.

Raised Panels Both Sides—Continued.

SIZE. Ft. In. × Ft. In.	Thickness. Inches.	Price First Quality. $ cts.	Price Second Quality. $ cts.	
2 6 × 6 6	1 3-4	5.25	4.60	Obey
2 8 × 6 8	"	5.50	4.75	Object
2 10 × 6 10	"	5.80	5.05	Obstruct
2 6 × 7 0	"	6.05	5.30	Overboil
2 8 × 7 0	"	6.10	5.35	Overcast
2 10 × 7 0	"	6.15	5.40	Overhaul
3 0 × 7 0	"	6.20	5.45	Occasion
2 6 × 7 6	"	6.80	6.05	Overflow
2 8 × 7 6	"	6.85	6.10	Obtain
2 10 × 7 6	"	6.95	6.20	Offend
3 0 × 7 6	"	7.00	6.25	Open
3 0 × 8 0	"	7.75	7.00	Oppress
3 0 × 8 6	"	8.80	8.05	Ordain
3 0 × 9 0	"	9.85	9.10	Order

Cipher is for 1st Quality. For 2d Quality add *ed.* 3d Quality add *ing.*

INCH DOORS.

Four-Panel O. G. Raised Panels One Side.

SIZE. Ft. In. × Ft. In.	Price First Quality. $ cts.	Price Second Quality. $ cts.	
2 0 × 6 0	1.80	1.65	Oust
2 4 × 6 4	1.90	1.75	Outpour
2 0 × 6 6	2.25	2.10	Own
2 6 × 6 6	2.25	2.10	Ornament
2 8 × 6 8	2.45	2.25	Offer

Inch Doors are made out of inch lumber, and finish up ⅞ in. thick.
Doors prepared for oil finish, 75c extra.
Sizes not given above, extra price.

Cipher is for 1st Quality. For 2d Quality add *d* or *ed*.

O G FIVE-PANEL DOORS—RAISED PANELS BOTH SIDES.

Cipher	SIZE. Ft.	In.		Ft.	In.	Thickness. Inches.	First Quality. $ Cts.	2nd Quality. $ Cts.
Quack	2	6	×	6	6	1 3-8	3.00	2.75
Quadrate	2	6	×	6	8	"	3.20	2.90
Quadruple	2	8	×	6	8	"	3.20	2.90
Quadruplicate	2	6	×	6	10	"	3.40	3.10
Quaff	2	8	×	6	10	"	3.45	3.15
Quail	2	10	×	6	10	"	3.50	3.20
Quake	2	0	×	7	0	"	3.60	3.30
Qualify	2	6	×	7	0	"	3.65	3.35
Quarrel	2	8	×	7	0	"	3.70	3.40
Quarry	2	10	×	7	0	"	3.75	3.45
Quarter	3	0	×	7	0	"	3.80	3.50
Quash	2	8	×	7	6	"	4.35	4.05
Quaver	2	10	×	7	6	"	4.40	4.10
Queen	3	0	×	7	6	"	4.45	4.15
Quell	3	0	×	8	0	"	5.10	4.80
Quench	3	0	×	8	6	"	6.00	5.70
Query	3	0	×	9	0	"	6.90	6.60
Quest	2	6	×	6	6	1 3-4	5.45	4.80
Question	2	8	×	6	8	"	5.70	4.95
Quill	2	10	×	6	10	"	6.00	5.25
Quibble	2	6	×	7	0	"	6.25	5.50
Quicken	2	8	×	7	0	"	6.30	5.55
Quiddle	2	10	×	7	0	"	6.35	5.60
Quilt	3	0	×	7	0	"	6.40	5.65
Quiet	2	6	×	7	6	"	7.00	6.25
Quirk	2	8	×	7	6	"	7.05	6.30
Quit	2	10	×	7	6	"	7 15	6.40
Quitclaim	3	0	×	7	6	"	7.20	6.45
Quiver	3	0	×	8	0	"	7.95	7.20
Quote	3	0	×	8	6	"	9.00	8.25
Quoit	3	0	×	9	0	"	10.05	9.30

P G FIVE-PANEL DOORS, RAISED PANELS BOTH SIDES.

Cipher is for 1st Quality. For 2d Quality add *d* or *ed*.

SIZE. Ft. In. × Ft In.	Thickness. Inches.	First Quality $ Cts.	2nd Quality $ Cts.	Cipher
2 6 × 6 6	1 3-8	3.20	2.95	Pace
2 6 × 6 8	"	3.40	3.10	Paddle
2 8 × 6 8	"	3.40	3.10	Page
2 6 × 6 10	"	3.60	3.30	Pain
2 8 × 6 10	"	3.65	3.35	Pale
2 10 × 6 10	"	3.70	3.40	Palliate
2 0 × 7 0	"	3.80	3.50	Palm
2 6 × 7 0	"	3.85	3.55	Palpitate
2 8 × 7 0	"	3.90	3.60	Palter
2 10 × 7 0	"	3 95	3.65	Pamper
3 0 × 7 0	"	4.00	3.70	Pander
2 8 × 7 6	"	4.55	4.25	Panel
2 10 × 7 6	"	4.60	4.30	Pant
3 0 × 7 6	"	4.65	4.35	Parade
3 0 × 8 0	"	5.30	5.00	Paragraph
3 0 × 8 6	"	6.20	5.90	Parallel
3 0 × 9 0	"	7.10	6.80	Paraphrase
2 6 × 6 6	1 3-4	5.65	5.00	Parboil
2 8 × 6 8	"	5.90	5.15	Parcel
2 10 × 6 10	"	6.20	5.45	Parch
2 6 × 7 0	"	6.45	5.70	Pardon
2 8 × 7 0	"	6.50	5.75	Pare
2 10 × 7 0	"	6.55	5.80	Park
3 0 × 7 0	"	6.60	5.85	Parole
2 6 × 7 6	"	7.20	6.45	Parry
2 8 × 7 6	"	7.25	6.50	Partake
2 10 × 7 6	"	7.35	6.60	Participate
3 0 × 7 6	"	7.40	6.65	Particularize
3 0 × 8 0	"	8.15	7.40	Partition
3 0 × 8 6	"	9.20	8.45	Paste
3 0 × 9 0	"	10.25	9.50	Pasture

Cipher is for F. M. 1 S. Square Top.
For F. M. 2 S. add *ed.*
Segment Top, prefix *un*
Circle Top, prefix *dis.*
1¾ inch add *ly.*

FOUR-PANEL MOULDED DOORS.

Flush or Sunk Mouldings.

Cipher	SIZE.				Thickness.	Price Moulded One Side.	Price Moulded two sides.
	Ft.	In.	Ft.	In.	Inches.	$ cts.	$ cts.
Rent	2	4 ×	6	4	1 3-8	3.95	4.65
Rack	2	0 ×	6	6	"	3.95	4.65
Roar	2	6 ×	6	6	"	3.95	4.65
Reach	2	6 ×	6	8	"	4.25	5.00
Rail	2	8 ×	6	8	"	4.25	5.00
Rest	2	10 ×	6	10	"	4.55	5.30
Revert	2	6 ×	7	0	"	4.90	5.65
Rivet	2	8 ×	7	0	"	4.95	5.70
Reward	2	10 ×	7	0	"	5.00	5.75
Revolt	3	0 ×	7	0	"	5.05	5.80
Rival	2	6 ×	7	6	"	5.65	6.40
Roast	2	8 ×	7	6	"	5.65	6.40
Rock	2	10 ×	7	6	"	5.70	6.45
Roll	3	0 ×	7	6	"	5.75	6.50
Remind	3	0 ×	8	0	"	6.45	7.25
Remember	2	8 ×	6	8	1 3-4	6.75	7.60
Remain	2	10 ×	6	10	"	7.15	8.00
Renew	3	0 ×	7	0	"	7.65	8.55
Render	2	8 ×	7	6	"	8.35	9.35
Repent	2	10 ×	7	6	"	8.45	9.45
Repair	3	0 ×	7	6	"	8.50	9.50
Repass	3	0 ×	8	0	"	9.30	10.30
Replenish	3	0 ×	8	6	"	10.25	11.25

Add to price of Moulded Doors for Circle Top Panels or Segment $1.45 for each side. Sizes not given above, extra price.

FOUR-PANEL MOULDED DOORS.

Raised Moulding.

Cipher is for R. M. 1 S. Square Top.
For R. M. 2 S. add *ed.*
Segment Top, prefix *un.*
Circle Top, prefix *dis.*
1¾ inch, add *ly.*

SIZE.	Thickness.	Price Moulded One Side.	Price Moulded two sides.	
Ft. In. Ft. In.	Inches.	$ cts.	$ cts.	
2 0 × 6 6	1 3-8	4 65	6.10	Reproach
2 6 × 6 6	"	4.65	6.10	Reprehend
2 6 × 6 8	"	5.00	6.45	Report
2 8 × 6 8	"	5.00	6.45	Repress
2 10 × 6 10	"	5.40	6.85	Represent
2 6 × 7 0	"	5.65	7.15	Republish
2 8 × 7 0	"	5.70	7.20	Resign
2 10 × 7 0	"	5.75	7.25	Resist
3 0 × 7 0	"	5.80	7.30	Respect
2 6 × 7 6	"	6.40	7.90	Resound
2 8 × 7 6	"	6.40	7.90	Respond
2 10 × 7 6	"	6.45	7.95	Retain
3 0 × 7 6	"	6.50	8.00	Result
3 0 × 8 0	"	7.25	8.75	Restrain
3 0 × 8 6	"	8.15	9.65	Ruin
2 8 × 6 8	1 3-4	7.60	9.05	Retreat
2 10 × 6 10	"	8.00	9.45	Retort
2 6 × 7 0	"	8.40	9.90	Return
2 8 × 7 0	"	8.45	9.95	Reveal
2 10 × 7 0	"	8.50	10.00	Review
3 0 × 7 0	"	8.55	10.05	Right
2 8 × 7 6	"	9.35	10.85	Riot
2 10 × 7 6	"	9.45	10.95	Roll
3 0 × 7 6	"	9.50	11.00	Root
3 0 × 8 0	"	10.30	11.80	Round
3 0 × 8 6	"	11.25	12.75	Rust

Add to price of Moulded Doors for Circle Top Panels or Segment $1.45 for each side. Sizes not given above, extra price.
All Doors prepared for Oil Finish, add 75c each extra.

O G SASH DOORS.

FIG. E.

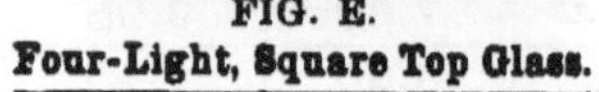

Four-Light, Square Top Glass.

FIG. F.

Two-Light, Circle Top Glass.

Cipher is for Unglazed.
For Glazed S. S. add *ed.*
Glazed D. S. add *edly.*
1¾ inch, prefix *un.*
For Two-Lights, prefix *re.*

Cipher	SIZE.	Thickness.	Four Lights, Square Top Glass. Price Unglaz'd	Four Lights, Square Top Glass. Price Glazed.	Two Lights, Circle Top Glass. Price Unglaz'd	Two Lights, Circle Top Glass. Price Glazed.
	Ft. In. Ft. In.	Inches.	$ cts.	$ cts.	$ cts.	$ cts.
Part	2 6 × 6 6	1 3-8	3.15	4.50	3.80	5.40
Pack	2 8 × 6 8	"	3.35	4.80	4.00	5.90
Paint	2 10 × 6 10	"	3.65	5.25	4.30	6.40
Patter	3 0 × 7 0	"	3.95	5.70	4.60	7.20
Pair	3 0 × 7 6	"	4.60	6.75	5.25	8.45
Pall	3 0 × 8 0	"	5.25	7.85	5.90	9.30

Doors for glass in first quality only. 1¾ Sash Doors same price as 1⅜

MOULDED SASH DOORS.

FIG. G. (Moulded One Side.). FIG. H.

Two Lights, Circle Top. **One Light, Upper Corners Circle.**

SIZE.	Thickness.	Two Lights, Circle Top Glass. Price Unglaz'd	Two Lights, Circle Top Glass. Price Glazed.	One Light, Upper Cor. Circle. Price Unglaz'd	One Light, Upper Cor. Circle. Price Glazed.	
Ft. In. Ft. In.	Inches.	$ cts.	$ cts.	$ cts.	$ cts.	
2 6 × 6 6	1 3-8	6.45	8.25	6.45	8.95	Pass
2 8 × 6 8	"	6.80	8.70	6.80	9.90	Pattern
2 10 × 6 10	"	7.20	9.35	7.20	10.60	Pawn
3 0 × 7 0	"	7.60	10.20	7.60	11.35	Peal
3 0 × 7 6	"	8.30	11.75	8.30	13.30	Peak
3 0 × 8 0	"	9.05	13.80	9.05	14.25	Peer

Cipher is for 2 L. C. T. R m 1 s., Unglazed.
For Glazed S. S. add *ed.*
Glazed D. S. add *edly.*
1¾ in. prefix *vn.*
R. M. 2 sides, prefix *dis.*
For 1 Light add

FOUR PANEL MACHINE CHAMFERED AND STUB MOULDED DOORS.

Size.	Price Cham. and End Moulded one side. 1⅜	Price Cham. and End Moulded two sides. 1⅜	Price Cham. and End Moulded one side. 1¾	Price Cham. and End Moulded two sides. 1¾
Ft. In. Ft. In	$ cts.	$ cts.	$ cts.	$ cts.
2 6×6 6	4.30	4.90	6.75	7.35
2 8×6 8	4.50	5.10	7.00	7 60
2 10×6 10	4.80	5.40	7.30	7.90
2 8×7 0	5.00	5.60	7.60	8.20
2 10×7 0	5.05	5.65	7.65	8.25
3 0×7 0	5.10	5.70	7.70	8.30
2 8×7 6	5.65	6.25	8.35	8.95
2 10×7 6	5.70	6.30	8.45	9.05
3 0×7 6	5.75	6.35	8.50	9.10
3 0×8 0	6.40	7.00	9.25	9.85
3 0×8 6	7.30	7.90	10.30	10.90

NO. 795—MACHINE CHAMFERED DOORS.

SIZE.	Thickness.	Price Chamfered One Side.	Price Chamfered Two Sides.
Ft. In. Ft. In.	Inches.	$ cts.	$ cts.
2 8 × 6 8	1 3-8	3.80	4.30
2 10 × 6 10	"	4.10	4.60
3 0 × 7 0	"	4.40	4.90
3 0 × 7 6	"	5.05	5.55
3 0 × 8 0	"	5.70	6.20

FIVE PANEL CHAMFERED AND STUB MOULDED DOORS.

SIZE.	Thickness.	Price Chamfered and Moulded One Side. O G One Side.	Price Chamfered and Moulded Two Sides.
Ft. In. Ft. In.	Inches.	$ cts.	$ cts.
2 8 × 6 8	1 3-8	5.95	8.15
2 10 × 6 10	"	6.35	8.55
3 0 × 7 0	"	6.75	9.00
3 0 × 7 6	"	7.45	9.70
3 0 × 8 0	"	8.20	10.45
2 8 × 6 8	1 3-4	8.55	10.75
2 10 × 6 10	"	8.95	11.15
3 0 × 7 0	"	9.50	11.75
3 0 × 7 6	"	10.45	12.70
3 0 × 8 0	"	11.25	13.50

NO. 732—FIVE PANEL MOULDED DOORS.

SIZE.	Thickness.	Price Moulded One Side.	Price Moulded Two Sides.
Ft. In. Ft. In.	Inches.	$ cts.	$ cts.
2 8 × 6 8	1 3-8	5.70	7.65
2 10 × 6 10	"	6.10	8.05
3 0 × 7 0	"	6.50	8.50
3 0 × 7 6	"	7.20	9.20
3 0 × 8 0	"	7.95	9.95

Cipher is for Unglazed.
For Glazed D. S. add *ed.*
S. D.—*Cipher* for Ungl'd.
For Glazed D. S. add *ed.*
With Shutters, add *edly.*
2¼, prefix *un.*
2½, prefix *dis.*
2¾, prefix *re.*

STORE DOORS.

Moulded Panel Outside—Single Strength Glass.

Code	SIZE.	Open per pair 1 3-8 inch.	Glazed Two-light each door per pair 1 3-8 inch.	Open per pair 1 3-4 inch.	Glazed Two-light each door per pair 1 3-4 inch.
	Ft. In. Ft. In.	$ cts.	$ cts.	$ cts.	$ cts.
Sack	4 6 × 7 0	9.75	12.45	15.20	17.90
Salt	4 6 × 7 6	10.50	13.80	16.90	20.20
Sand	4 6 × 8 0	11.40	14.75	18.60	21.95
School	5 0 × 7 0	10.40	13.70	15.75	19.05
Scoff	5 0 × 7 6	11.40	15.25	17.50	21.35
Scoop	5 0 × 8 0	12.50	16.90	19.15	23.55
Scout	5 0 × 8 6	13.65	18.85	21.00	26.20

Add for Double Strength Glass 10 per cent of 1¾ Glazed List.
Double Thick, add 50 to 100 per cent. to Open List.

All Store Doors are made without Shutters, unless otherwise ordered. We also make Store Doors and Store Sash, moulded both sides on the glass, either square, circle-top, segment, or circle-corners, to order. In ordering, always state whether glazed or unglazed, and whether single or double strength glass is wanted, and how many lights in each door.

Cipher is for Open Sash.
For Glazed S. S. add *ed.*
Glazed D. S, add *edly.*
1¾ in. prefix *un.*

DOUBLE FRONT DOORS.

Heavy Raised Mouldings Outside, Circle Top Panels.

Code	SIZE.	Thickness.	Price per Pair.
	Ft. In. Ft. In.	Inches.	$ cts.
Spank	4 0 × 7 0	1 3-4	20.70
Spark	4 4 × 7 0	"	20.70
Spatter	4 6 × 7 0	"	20.70
Spawl	4 6 × 7 6	"	22.60
Spawn	4 6 × 8 0	"	24.50
Spay	5 0 × 7 6	"	22.60
Spear	5 0 × 8 0	"	24.50

COTTAGE FRONT DOORS.

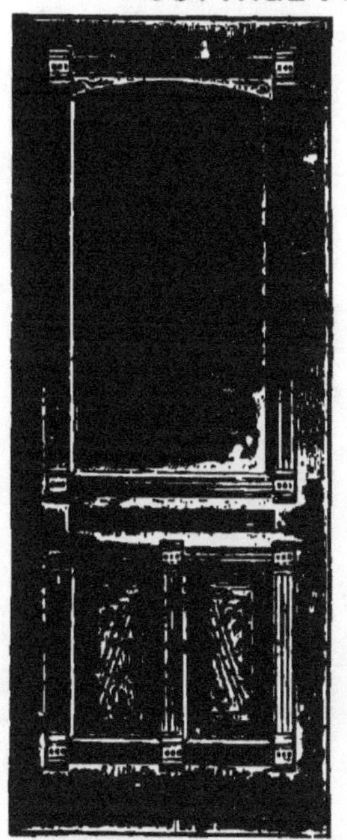

FIG. K.
GRANT.

FIG. L.
OSHKOSH.

Size.	Thickness.	GRANT—1-LT. Price Open.	GRANT—1-LT. Price Glazed.	OSHKOSH—1-LT. Price Open.	OSHKOSH—1-LT. Price Glazed.
Ft. In. Ft. In.	Inch's.	$ cts.	$ cts.	$ cts.	$ cts.
2 6 × 6 6	1 3-8	6.75	9.10	9.00	11.35
2 8 × 6 8	"	7.10	10.10	9.30	12.30
2 10 × 6 10	"	7.50	10.85	9.65	13.00
3 0 × 7 0	"	7.90	11.70	10.10	13.90
3 0 × 7 6	"	8.60	13.40	10.75	15.55
3 0 × 8 0	"	9.35	14.55	11.65	16.85

For 1¾ Doors add to this list the difference between 1⅜ and 1¾ Four-Panel Doors, same size.

N. B.—We now make these doors with a 12¼ inch Lock Rail.

COTTAGE FRONT DOORS.

INDIANA.

LEE.

Size.	Indiana Thickness.	Indiana Price Unglaz'd	Indiana Price Glazed.	Lee Thickness.	Lee Price Unglaz'd	Lee Price Glazed.
Ft. In. Ft. In.	Inch's.	$ cts.	$ cts.	Inch's.	$ cts.	$ cts.
2 6 x 6 6	1 3-8	5.20	7 55	1 3-4	....	....
2 8 x 6 8	"	5 55	8.55	"	....	.. ..
2 10 x 6 10	"	5.95	9.30	"	....	....
3 0 x 7 0	"	6.35	10.15	"	21.00	24.55
3 0 x 7 6	"	7.05	11.85	"	22.00	25 85
3 0 x 7 0	"	7.80	13.00	"	23.00	28 00

N. B.—We make the Lee Door with a wide Lock Rail.

COTTAGE FRONT DOORS.

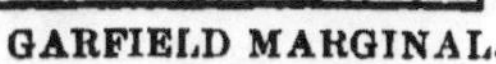
GARFIELD MARGINAL.

ILLINOIS MARGINAL.

Net prices of the above doors furnished upon application.

We now make these doors with any style of marginal light and if other than what is shown above is desired, indicate by sketch.

N. B.—We make these doors with a 12¼ inch lock rail.

COTTAGE FRONT DOORS.

OSHKOSH MARGINAL. GRANT MARGINAL.

Net prices of the above doors furnished upon application.

We make these doors with any style of marginal light, and if other than what is shown above is desired, indicate by sketch.

N. B.—We now make these doors with a 12¼ inch Lock Rail.

COTTAGE FRONT DOORS.

OSHKOSH PANEL.

GARFIELD PANEL.

Net prices of the above doors furnished upon application.

We make these doors with or without carving on panel, as desired.

We also make other styles of the Cottage Doors, with panels, if desired.

N. B.—We now make these doors with a 12½ inch Lock Rail.

COTTAGE FRONT DOORS.

OSHKOSH FRONT AND VESTIBULE DOORS.

Net prices of the above doors furnished upon application.
We make these doors in pairs, filling an opening of 4 0 x 7 0 to 5 0 x 8 6
Panels of the Front Doors may be either plain or carved, as desired.
N. B.—We make these doors with a 12½ inch lock rail.

COTTAGE FRONT DOORS.

CLEVELAND.

QUEEN ANNE.

Net prices of the above doors furnished upon applination.

We glaze the sash doors of the Cottage Front Door series with either plain or fancy glass. The marginal lights are glazed with cathedral glass if so desired.

For glass suitable for these doors see pages 94-99 inclusive.

COTTAGE FRONT DOORS.

NASHVILLE. GARFIELD—2-LIGHT.

Net prices of the above doors furnished upon application.
We make the Nashville Door without marginal lights, if so desired.

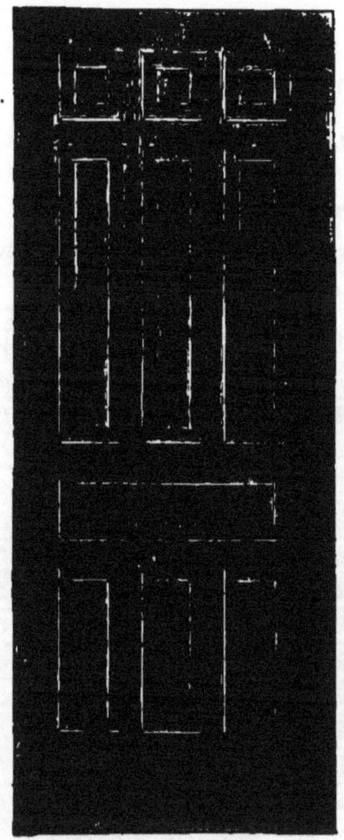

10 Panel O G Door.

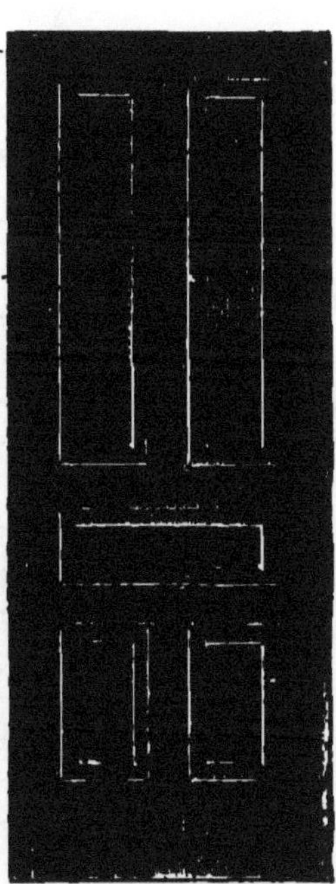

5 Panel O G Door.

We are making a specialty of furnishing doors, ordinary sizes, with 5, 6, 7, 9 and 10 panels. The 6 panel door is made 3 panels wide and 2 panels high, as well as 2 panels wide and three high, the top panels being square like those of the 10 panel door. The 7 panel door is like the 10 panel with the three square top panels left off. The 9 panel door has 3 panels wide and 3 panels high. Please designate clearly in ordering just what is wanted. These doors can be made with P G sticking as well as O G.

- *Cipher* is for 1⅛ thick.
- For 1⅜ inches add *ed.*
- Segment Top, prefix *un.*
- Circle Top, prefix *dis.*
- Stationary Slats, ad *ly.*
- Half Stationary, add *ism.*

OUTSIDE BLINDS.

Twelve-Lighted Windows.

	SIZE.	Price Rolling Slats. 1 1-8 inch.	Price for Stationary or ½ St. Slats 1 1-8 inch.	Price for Stationary or ½ St. Slats 1 3-8 inch.	Size of Blinds.
	Inches.	$ cts.	$ cst.	$ cts.	Ft. In. Ft. In.
Sequester	8 × 10	1.55	1.80	2.30	2 4½ × 3 11
Shadow	8 × 12	1.85	2.10	2.60	2 4½ × 4 7
Shuck	8 × 14	2.00	2.25	2.75	2 4½ × 5 3
Shaft	9 × 12	1.85	2.10	2.60	2 7½ × 4 7
Sharpen	9 × 13	2.00	2.25	2.75	2 7½ × 4 11
Shatter	9 × 14	2.00	2.25	2.75	2 7½ × 5 3
Shear	9 × 15	2.20	2.45	2.95	2 7½ × 5 7
Shelter	9 × 16	2.20	2.45	2.95	2 7½ × 5 11
Shimmer	9 × 18	2.50	2.75	3.25	2 7½ × 6 7
Shield	10 × 12	2.00	2.25	2.75	2 10½ × 4 7
Shock	10 × 14	2.20	2.45	2.95	2 10½ × 5 3
Shipwreck	10 × 15	2.35	2.60	3.10	2 10½ × 5 7
Shoal	10 × 16	2.35	2.60	3.10	2 10½ × 5 11
Shout	10 × 18	2.65	2.90	3.40	2 10½ × 6 7
Shoulder	10 × 20	2.95	3.20	3.70	2 10½ × 7 3
Show	10 × 22	3.25	3.50	4.00	2 10½ × 7 11
Shove	10 × 24	3.50	3.75	4.25	2 10½ × 8 7

1⅜ thick, add to price of 1⅛, per window, 50 cts. Segment Head Blinds, add 75 cts. per pair. Half Circle Head Blinds, add $1.25 per pair. Size of Blinds measure same as Check Rail Windows, with the addition of 1 inch to the bottom rail, for Sub-sill Window Frames, which can be cut off if necessary.

For 12-inch Twelve-Light Blinds, add 35c to price of 10-inch.

For 15 inch Four-Light and 30 inch Two-Light Blinds, see price of 10 inch Twelve-Light same height.

OUTSIDE BLINDS.

Eight-Lighted Windows.

Cipher, for Rolling Slats.
Stationary Slats, add *ed*.
Half Stationary, add *edly*.
15 Light, prefix *un*.
10 Light, prefix *dis*.

SIZE.	Price Rolling Slats. 1 1-8 inch.	Price for Stationary or ½ St. Slats 1 1-8 inch.	Price for Stationary or ½ St. Slats 1 3-8 inch.	Size of Blinds.	
Inches.	$ cts	$ cts.	$ cts.	Ft. In. Ft. In.	
9 × 12	1.85	2.10	2.60	1 11 × 4 7	Shower
9 × 14	2.00	2.25	2.75	1 11 × 5 3	Shunt
9 × 16	2.20	2.45	2.95	1 11 × 5 11	Shroud
9 × 18	2.50	2.75	3.25	1 11 × 6 7	Sight
10 × 12	1.85	2.10	2.60	2 1 × 4 7	Signal
10 × 14	2.00	2.25	2.75	2 1 × 5 3	Silver
10 × 16	2.20	2.45	2.95	2 1 × 5 11	Simmer
10 × 18	2.50	2.75	3.25	2 1 × 6 7	Sketch
10 × 20	2.75	3.00	3.50	2 1 × 7 3	Skill
12 × 14	2.00	2.25	2.75	2 5 × 5 3	Skunk
12 × 16	2.20	2.45	2.95	2 5 × 5 11	Slash
12 × 18	2.50	2.75	3.25	2 5 × 6 7	Slack
12 × 20	2.75	3.00	3.50	2 5 × 7 3	Slander

DOOR BLINDS.

SIZE.	Thickness.	Price.	
Ft. In. Ft. In.	Inches.	$ cts.	
2 6 × 6 6	1 1-8	2.50	Storm
2 8 × 6 8	"	2.75	Strain
2 10 × 6 10	"	3.00	Streak
3 0 × 7 0	"	3.30	Stretch

1⅜ thick, add to price of 1⅛, 50c per pair.
In ordering, state if Blinds are to be a pair or single piece, to each opening.
Sizes not given extra price.

OUTSIDE BLINDS.

Four-Lighted Windows.

Cipher is for 1⅛ thick. For 1⅜ inch, add *ed.* Segment Top, prefix *un.* Circle Top, prefix *dis.* Stationary Slats, add *ly.* Half Stat'y Slats, add *ism*

	SIZE.	Price Rolling Slats 1 1-8 inch.	Price for Stationary or ½ St. Slats 1 1-8 inch.	Price for Stationary or ½ St. Slats 1 3-8 inch.	Size of Blinds.			
	Inches.	$ cts.	$ cts.	$ cts.	Ft.	In.	Ft.	In.
Sliver	12 × 20	1.55	1.80	2.30	2	5 ×	3	11
Slaughter	12 × 22	1.85	2.10	2.60	2	5 ×	4	3
Slough	12 × 24	1.85	2.10	2.60	2	5 ×	4	7
Slump	12 × 26	2.00	2.25	2.75	2	5 ×	4	11
Slumber	12 × 28	2.00	2.25	2.75	2	5 ×	5	3
Smack	12 × 30	2.20	2.45	2.95	2	5 ×	5	7
Smart	12 × 32	2.20	2.45	2.95	2	5 ×	5	11
Smelt	12 × 34	2.50	2.75	3.25	2	5 ×	6	3
Smatter	12 × 36	2.50	2.75	3.25	2	5 ×	6	7
Smooth	12 × 38	2.75	3.00	3.50	2	5 ×	6	11
Snort	12 × 40	2.75	3.00	3.50	2	5 ×	7	3
Snatch	14 × 24	2.00	2.25	2.75	2	9 ×	4	7
Sneak	14 × 26	2.20	2.45	2.95	2	9 ×	4	11
Snood	14 × 28	2.20	2.45	2.95	2	9 ×	5	3
Snuff	14 × 30	2.35	2.60	3.10	2	9 ×	5	7
Sneer	14 × 32	2.35	2.60	3.10	2	9 ×	5	11
Snivel	14 × 34	2.65	2.90	3.40	2	9 ×	6	3
Soak	14 × 36	2.65	2.90	3.40	2	9 ×	6	7
Soften	14 × 38	2.95	3.20	3.70	2	9 ×	6	11
Soar	14 × 40	2.95	3.20	3.70	2	9 ×	7	3
Sober	14 × 42	3.25	3.50	4.00	2	9 ×	7	7
Soil	14 × 44	3.25	3.50	4.00	2	9 ×	7	11
Soldier	14 × 46	3.50	3.75	4.25	2	9 ×	8	3
Sort	14 × 48	3.50	3.75	4.25	2	9 ×	8	7

Outside Blinds—Two-Lighted Windows.

Blinds for Two-Lighted Windows, with glass 20, 22 or 24 inches wide, same price as Blinds for Four-Lighted Windows with 12-in. glass, same height.

Blinds for Two-Lighted Windows, with glass 26 or 28 inches wide, same price as Blinds for Four-Lighted Windows with 14-in. glass, same height.

OUTSIDE BLINDS.

Two-Lighted Pantry Windows.

Size of Glass.	Thickness.	Price.	Size of Blinds.
Inches.	Inches.	$ cts.	Ft. In. Ft. In.
12 × 24	1 1-8	.95	1 4⅛ × 4 7½
12 × 26	"	1.00	1 4⅛ × 4 11½
12 × 28	"	1.00	1 4⅛ × 5 3½
12 × 30	"	1.10	1 4⅛ × 5 7½
12 × 32	"	1.10	1 4⅛ × 5 11½
12 × 34	"	1.25	1 4⅛ × 6 3½
12 × 36	"	1.25	1 4⅛ × 6 7½
14 × 24	"	1.00	1 6⅛ × 4 7½
14 × 26	"	1.00	1 6⅛ × 4 11½
14 × 28	"	1.10	1 6⅛ × 5 3½
14 × 30	"	1.20	1 6⅛ × 5 7½
14 × 32	"	1.20	1 6⅛ × 5 11½
14 × 34	"	1.35	1 6⅛ × 6 3½
14 × 36	"	1.35	1 6⅛ × 6 7½
14 × 40	"	1.50	1 6⅛ × 7 3½

Four-Lighted Pantry Windows.

Size of Glass.	Thickness.	Price	Size of Blinds.
Inches.	Inches.	$ cts.	Ft. In. Ft. In.
8 × 10	1 1-8	.80	1 0 × 3 11½
9 × 12	"	.95	1 1 × 4 7½
9 × 14	"	1.00	1 1 × 5 3½
9 × 16	"	1.10	1 1 × 5 11½
9 × 18	"	1.25	1 1 × 6 7½
10 × 12	"	.95	1 2 × 4 7½
10 × 14	"	1.00	1 2 × 5 3½
10 × 16	"	1.10	1 2 × 5 11½
10 × 18	"	1.25	1 2 × 6 7½

For 12 and 14 inch, see Two- Light List above, same height.

INSIDE BLINDS.

Fig. 1—All Slats. **Fig. 2—½ Panels, ½ Slats.**

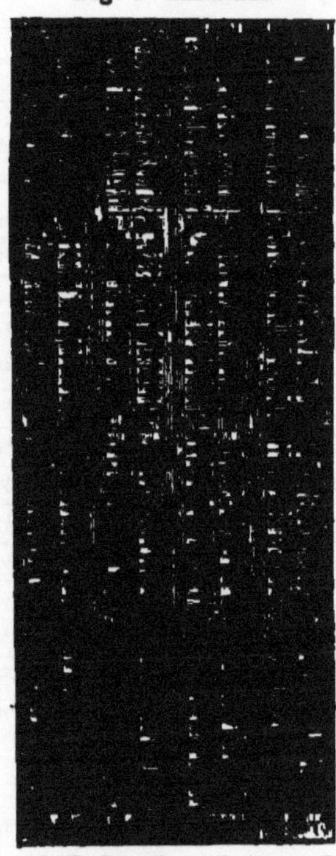

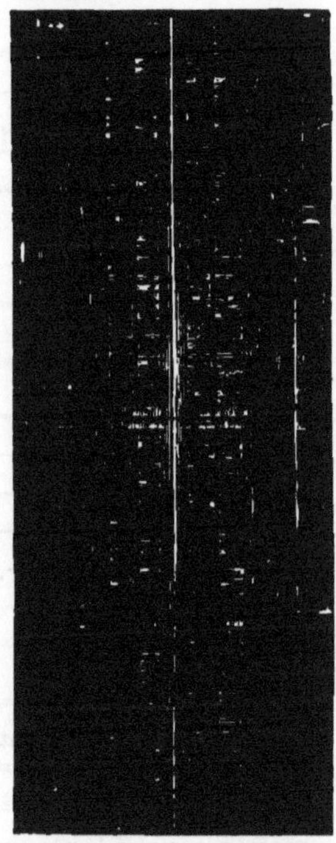

O G Panel or Rolling Slats, measuring height of window, ordinary width, per foot:

Two-fold	$ .70	Four-fold	$1.10
Three-fold	.90	Six-fold	1.50

The above prices are for pine. If hard wood, such as Cherry, Ash, Maple or Black Walnut, we charge about double the price of Pine. We make Inside Blinds that are not excelled, either in workmanship or style, in this market.

DIRECTIONS FOR ORDERING.

First—In all cases give the exact outside measure of Blinds wanted.

Second—Give the number of folds.

Third—State if Blinds are to be all slats, or ½ panels and ½ slats.

Fourth—State distance from top of window to center of meeting rail or sash, or where Blinds are to be cut.

Fifth—Give thickness of Blinds. They are made 1⅛ inch thick unless otherwise ordered.

Sixth—If Blinds fold in pockets, give the size of pockets.

Seventh—State if Blinds are to be painted or finished in oil.

SECTIONS OF DOOR AND WINDOW FRAMES.

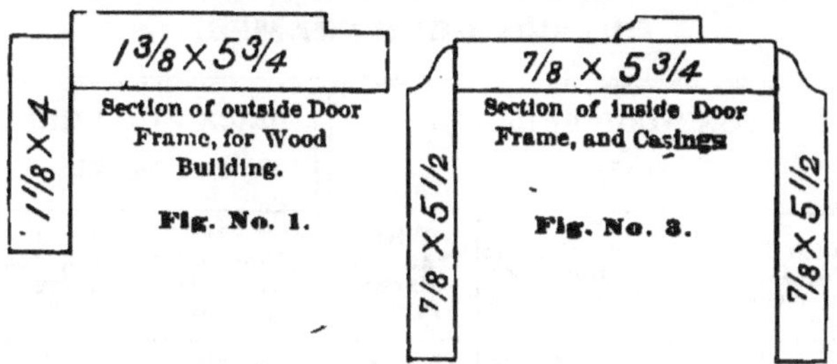

Section of outside Door Frame, for Wood Building.

Fig. No. 1.

Section of inside Door Frame, and Casings

Fig. No. 3.

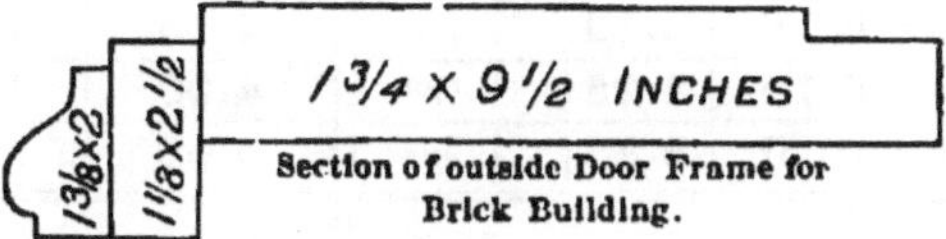

Section of outside Door Frame for Brick Building.

Fig. No. 2.

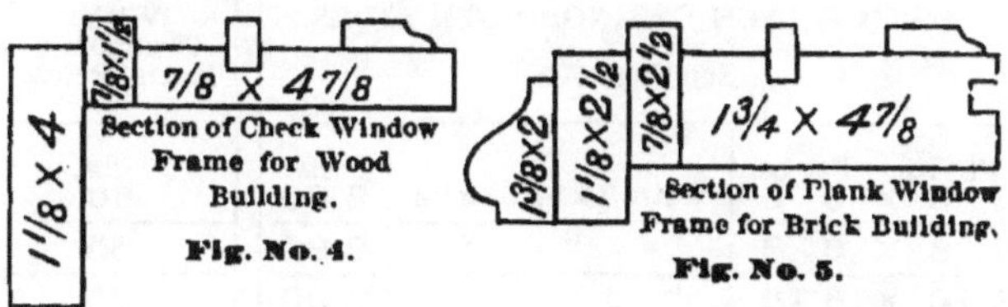

Section of Check Window Frame for Wood Building.

Fig. No. 4.

Section of Plank Window Frame for Brick Building.

Fig. No. 5.

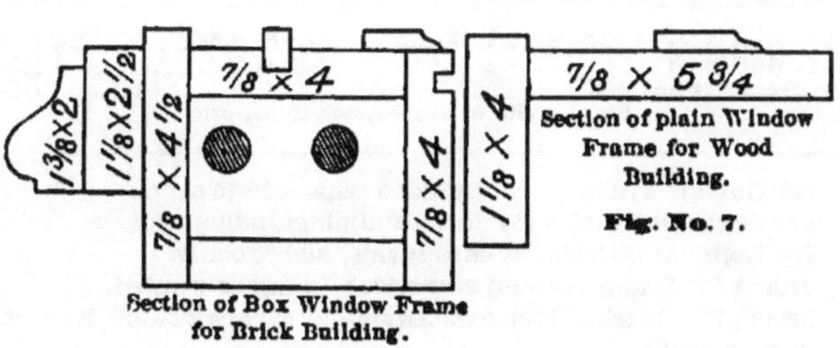

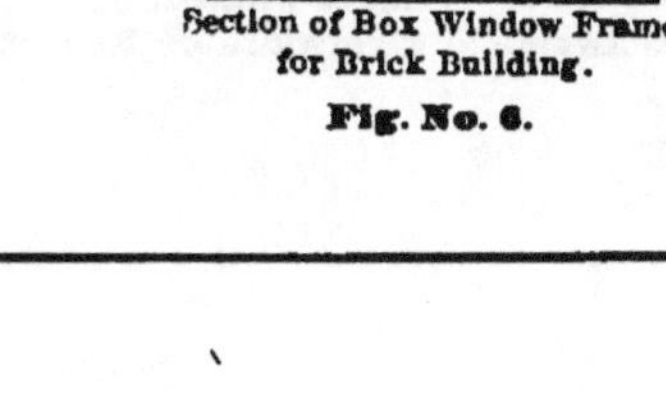

Section of plain Window Frame for Wood Building.

Fig. No. 7.

Section of Box Window Frame for Brick Building.

Fig. No. 6.

OUTSIDE DOOR FRAMES.

FOR WOOD BUILDING. See Page 57.			BRICK BUILDING. S ceFage 57.	
With Plain Outside Casings 1⅛×4⅛ in. and 1⅜ in Rabbeted Jambs, for 5 inch Stud.		With Transoms Extra Price.	With 1 3-4 inch Rabbeted Jambs and Moulded Outside Casing.	With Transoms Extra Price.
Ft. In. Ft. In.	$ cts.	$ cts.	$ cts.	$ cts.
2 6 × 6 6	3.00	80	4.65	90
2 8 × 6 8	3.10	90	4.75	1.00
2 10 × 6 10	3.25	1.00	4.90	1.10
3 0 × 7 0	3.35	1.10	5.00	1.20

INSIDE DOOR FRAMES.

WITH 5 INCH CASINGS BOTH SIDES. See Page 57.			With Transoms, Extra Price.
Ft. In. Ft. In.		$ cts.	$ cts.
2 6 × 6 6	With ⅞×5¾ Jambs.	2.80	.60
2 8 × 6 8	" " "	2.90	.65
2 10 × 6 10	" " "	3.00	.75
3 0 × 7 0	" " "	3.10	.80

2 ft. 6x6 ft. 6 Inside Frames, Jambs 1 3-4x5 7-8..........$1.50
2 ft. 6x6 ft. 8 " " " 7-8x5 7-8.......... 1.20
Transom Frames, add.. .50

Extra width of Jambs, additional price.

For Outside Frames with Moulded Caps, add 75 cents.

For Segment Frames for Brick Buildings, add $1.50.

For Segment Outside, Square Inside, add 60 cents.

Orders for Frames should give width amba wanted.

Orders for Outside Door Frames should give width of Rabbet to receive door.

Extra Price for wider Jambs.

WINDOW FRAMES.

Four-Lighted Windows.

FOR WOOD BUILDING.		FOR BRICK BUILDING.	
For Check Rail Sash, Outside Casing 1⅛×4½ inches. See Page 57.		1¾ in. Plank, with Moulding. See Page 57.	Box Frames, with Moulding. See Page 57.
SIZES OF GLASS. In. In. In. × In.	PRICE. $ cts.	PRICE. $ cts.	PRICE. $ cts.
10 12 14 × 20	2.40	2.75	3.60
10 12 14 × 22	2.50	3.00	4.00
10 12 14 × 24	2.65	3.25	4.30
10 12 14 × 26	2.85	3.40	4.45
10 12 14 × 28	3.00	3.60	4.55
10 12 14 × 30	3.10	3.75	4.70
10 12 14 × 32	3.15	3.85	4.80
10 12 14 × 34	3.25	4.00	5.00
10 12 14 × 36	3.40	4.15	5.20
10 12 14 × 40	3.65	4.35	5.50
10 12 14 × 44	3.90	4.60	5.75
10 12 14 × 48	4.20	4.80	6.00

For Circle Top or Circle Corner Window or Door Frames add to List Prices of Square Top $3.00.

Frames with inside finish add to List Prices $1.00.

Frames with pulleys for weights add 75 cents.

Segment outside, square inside, add 60 cents.

Frames for Two-Light Windows same price as Four-Light, same length.

Frames with Moulded Cap, add 75 cents.

WINDOW FRAMES.

Twelve-Lighted Windows.

SIZE. OF GLASS.	WOOD BUILDING.		BRICK BUILDING.	
	Plain Rail Sash. See Page 57.	Check Rail Sash. See Page 57.	1¾ in. Plank, and Moulding. See Page 57.	Box Frames, and Moulding. See Page 57.
Inches.	$ cts.	$ cts.	$ cts.	$ cts.
8 × 10	1.65	2.40	2.75	3.60
9 × 12	1.90	2.65	3.25	4.30
9 × 14	2.15	3.00	3.60	4.55
9 × 15	2.40	3.10	3.75	4.70
9 × 16	2.40	3.15	3.85	4.80
9 × 18	2.65	3.40	4.15	5.20
10 × 12	2.00	2.80	3.40	4.45
10 × 14	2.25	3.15	3.75	4.70
10 × 16	2.50	3.30	4.00	4.95
10 × 18	2.75	3.55	4.30	5.35
10 × 20	3.00	3.80	4.50	5.65

Eight-Lighted Windows.

SIZE OF GLASS.	WOOD BUILDING.		BRICK BUILDING.	
	Plain Rail Sash. See Page 57.	Check Rail Sash. See Page 57.	1¾ in. Plank, and Moulding. See Page 57.	Box Frames, and Moulding. See Page 57.
Inches.	$ cts.	$ cts.	$ cts.	$ cts.
12 × 14	2.15	3.00	3.60	4.55
12 × 16	2 40	3.15	3.85	4.80
12 × 18	2 65	3.40	4.15	5.20
12 × 20	2.90	3.65	4.35	5.50
12 × 22	3.15	3.90	4.60	5.75
12 × 24	3.40	4.20	4.80	6.00

WEICHTS.

BLINDS.

Four-Light Windows.

SIZE.	Thickness.	Weight
12 × 20	1 1-8	14 lbs.
12 × 24	"	16 "
12 × 28	"	18 "
12 × 32	"	20 "
12 × 36	"	23 "
12 × 40	"	25 "
12 × 44	"	28 "

Eight-Light Windows.

SIZE.	Thickness.	Weight
9 × 12	1 1-8	15 lbs.
9 × 16	"	18 "
10 × 14	"	18 "
12 × 14	"	20 "
12 × 16	"	23 "
12 × 20	"	26 "

Twelve-Light Windows.

SIZE.	Thickness.	Weight
8 × 10	1 1-8	14 lbs.
9 × 12	"	18 "
9 × 16	"	23 "
10 × 14	"	22 "
10 × 18	"	27 "
10 × 20	"	30 "

WINDOWS.

Two-Light Check Rail.

SIZE.	Thickness.	Glazed Single Strength	Glazed Double Strength
20 × 24	1 3-8	21 lbs.	23 lbs.
20 × 28	"	22 "	25 "
20 × 32	"	23 "	26 "
20 × 36	"	25 "	28 "
20 × 40	"	26 "	30 "
24 × 30	"	24 "	26 "
24 × 32	"	25 "	28 "
24 × 36	"	27 "	30 "
24 × 40	"	29 "	33 "
26 × 30	"	25 "	28 "
26 × 32	"	26 "	30 "
26 × 34	"	27 "	31 "
26 × 36	"	28 "	32 "
26 × 40	"	30 "	34 "
26 × 44	"	32 "	36 "
26 × 48	"	34 "	39 "
28 × 32	"	28 "	32 "
28 × 36	"	30 "	34 "
28 × 40	"	32 "	36 "
28 × 44	"	34 "	38 "
28 × 48	"	36 "	40 "

Weight of Four-Light Windows is practically the same as Two-Light of same length.

WEIGHTS.

Eight-Lighted Windows.

Windows—Plain Rail Sash.			Windows—Check Rail Sash.		
SIZE.	Weight Glazed.	Weight Unglaz'd	SIZE.	Weight Glazed.	Weight Unglaz'd
8 × 10	10 lbs.	5 lbs.	9 × 12	17 lbs.	8 lbs.
8 × 12	12 "	5 "	9 × 14	18 "	9 "
8 × 14	14 "	7 "	10 × 12	18 "	9 "
9 × 12	14 "	6 "	10 × 14	19 "	11 "
9 × 14	17 "	7 "	10 × 16	22 "	12 "
10 × 12	15 "	8 "	12 × 14	23 "	11 "
10 × 14	18 "	8 "	12 × 16	24 "	12 "
10 × 16	20 "	9 "	12 × 18	27 "	13 "
12 × 14	19 "	9 "	12 × 20	32 "	14 "
12 × 16	22 "	11 "	14 × 20	35 "	15 "
12 × 18	25 "	12 "	14 × 24	40 "	17 "

Twelve-Lighted Windows.

Plain Rail Sash.			Check Rail Sash.		
SIZE.	Weight Glazed.	Weight Unglaz'd	SIZE.	Weight Glazed.	Weight Unglaz'd
8 × 10	14 lbs.	6 lbs	8 × 12	20 lbs.	8 lbs.
8 × 12	18 "	8 "	8 × 14	22 "	8 "
8 × 14	19 "	8 "	9 × 12	22 "	9 "
9 × 12	20 "	9 "	9 × 14	24 "	10 "
9 × 14	22 "	9 "	9 × 16	27 "	11 "
9 × 16	26 "	9 "	10 × 12	23 "	11 "
10 × 12	21 "	9 "	10 × 14	26 "	11 "
10 × 14	23 "	9 "	10 × 16	29 "	12 "
10 × 16	26 "	10 "	10 × 18	32 "	13 "
12 × 14	25 "	10 "	10 × 20	34 "	14 "
12 × 16	28 "	10 "	12 × 20	36 "	14 "
12 × 18	31 "	10 "	12 × 24	42 "	15 "

WEIGHTS.

Four-Panel Doors.

SIZE.	THICKNESS.			
Ft. In. Ft. In.	1 Inch.	1⅛ Inch.	1⅜ Inch.	1¾ Inch.
2 0 × 6 0	17 lbs.	22 lbs.		
2 4 × 6 4	21 "	26 "		
2 6 × 6 6	23 "	28 "	33 lbs.	40 lbs.
2 8 × 6 8	24 "	30 "	35 "	44 "
2 10 × 6 10		33 "	37 "	47 "
3 0 × 7 0	•	35 "	40 "	49 "
3 0 × 7 6			42 "	53 "

For Moulded Doors add to above five pounds for each side moulded.

ESTIMATED WEIGHT OF LUMBER, ETC., DRY.

Flooring, Dressed and Matched, per 1,000 feet	....	1,800 lbs.
Poplar Box Boards,	" "	2,000 "
Siding, Dressed,	" "	800 "
Ceiling, ⅜ inch thick,	" "	800 "
" ½ " "	" "	900 "
Boards, Dressed one side,	" "	2,000 "
" and Dimension, rough,	" "	2,400 "
Shingles,	" pieces	240 "
Lath,	" "	500 "
Pickets, Dressed,	" "	1,800 "
" Rough,	" "	2,400 "

WEIGHT OF MOULDINGS.

1×1 inch per one hundred lineal feet, fifteen pounds.

STAIRS,

STAIR RAILING, BALUSTERS,

NEWEL POSTS, ETC.

STAIR RAILING

Of every description, made ready to put up, with crooks, etc. worked and finished in the best manner, of Pine, Butternut Ash, Oak, Black Walnut, Cherry or Mahogany.

We keep constantly on hand, or make to order, all kinds of Newel Posts and Balusters.

PRICES LOW. ESTIMATES OR INFORMATION GIVEN ON APPLICATION.

ORDERS FOR STAIRS

Should give the height of story from floor to floor, width of joist in second story, width and run of stairs and size of cylinder, with sketch, showing something of the shape wanted, and the way they turn on the landing; also style and width of base used in hall.

ORDERS FOR RAILING

Worked for straight flight, should give the rise and tread of steps as sawed out on the string board, the number of risers, the size of cylinder, which way it turns on landing, and length of straight rail required at head of stairs.

FOR WINDING STAIRS

We should have an exact plan, showing location of risers in cylinder, the width of steps, risers, etc.

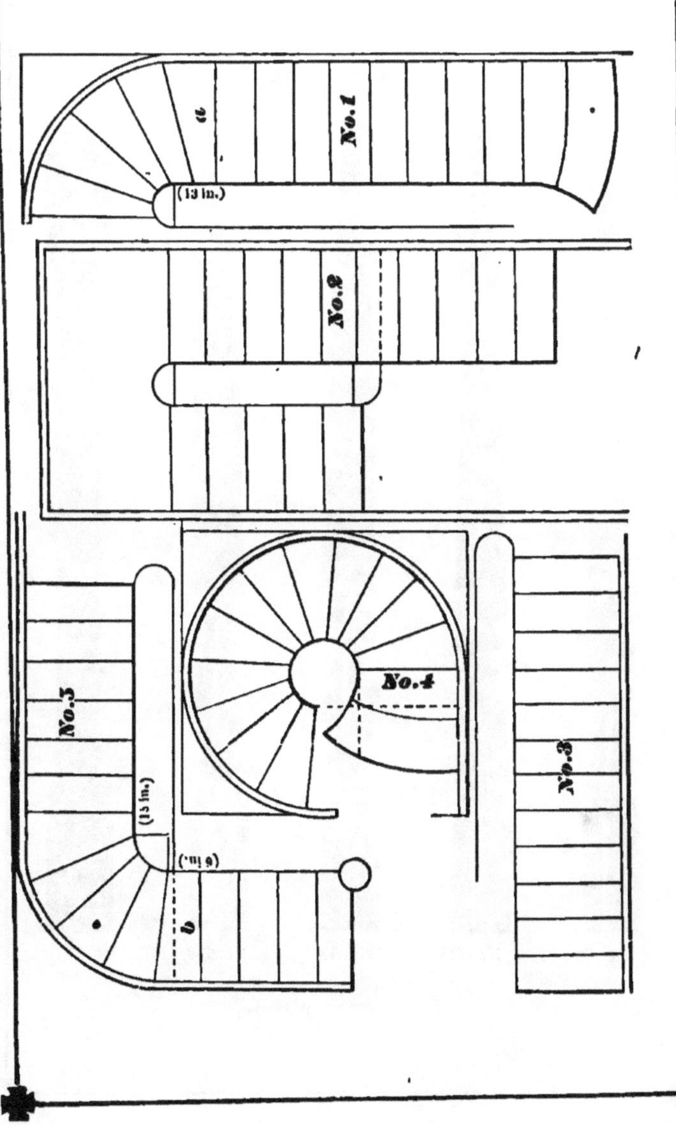
No.1
a
(13 in.)
No.2
No.3
(15 in.)
(6 in.)
b
No.4
No.3

BALUSTERS.

No. 782. No. 783. No. 784. No. 785. No. 787.

For prices see page 76.

BALUSTERS.

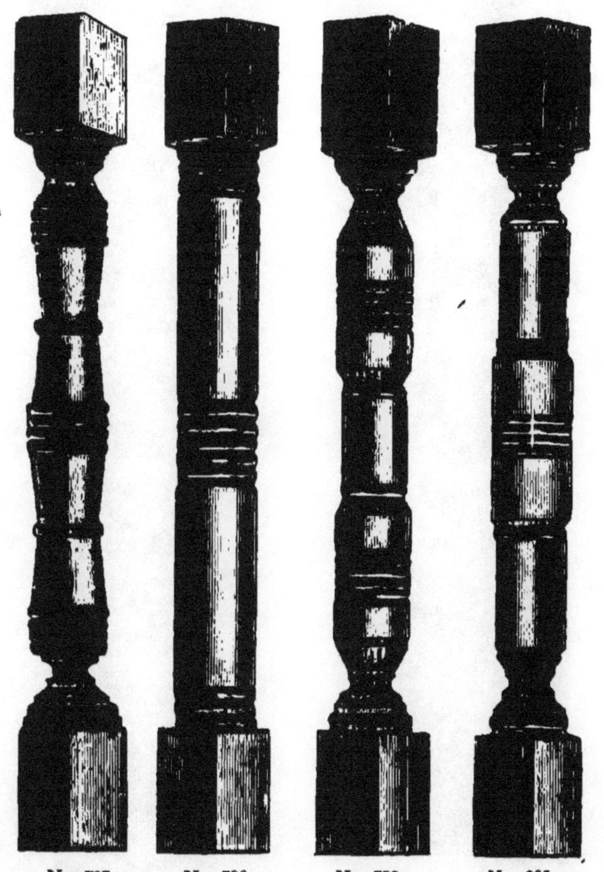

No. 797. No. 798. No. 799. No. 800.

No 761.

No. 762.

For prices see page 76.

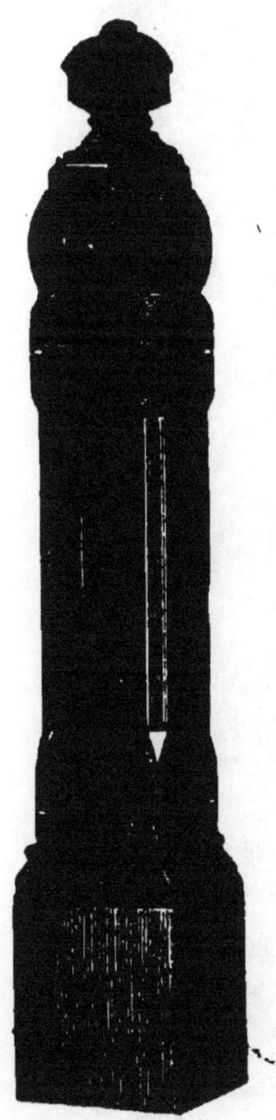

No. 5.

No. 763.

For prices see page 76.

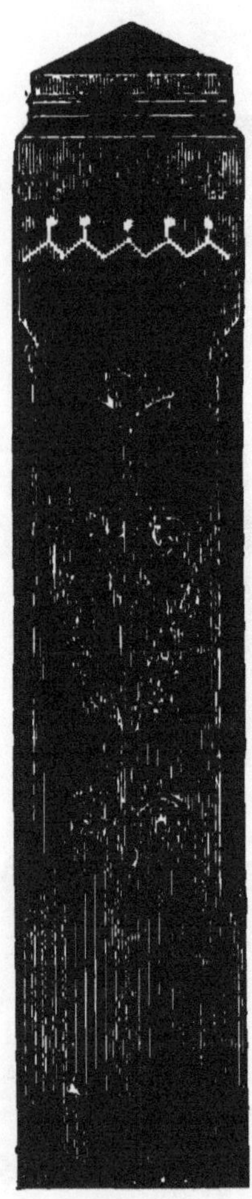

No. 777.

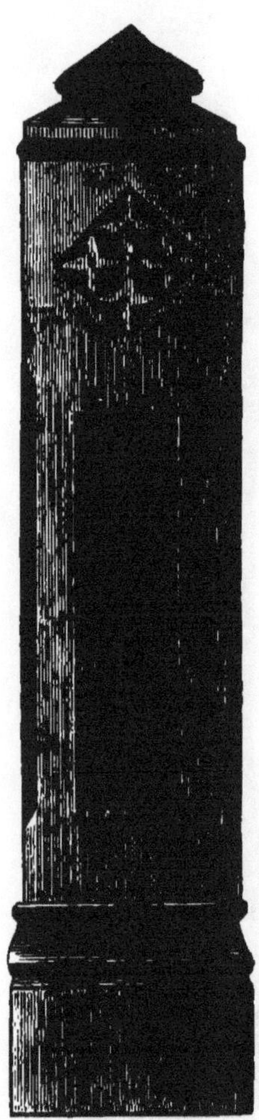

No. 778.

For prices see page 76.

No. 7.

No. 8.

For prices see page 76.

ANGLE NEWELS FOR PLATFORM STAIRS.

No. 1. No. 2. No. 3.

For prices see page 76.

STAIR RAIL.

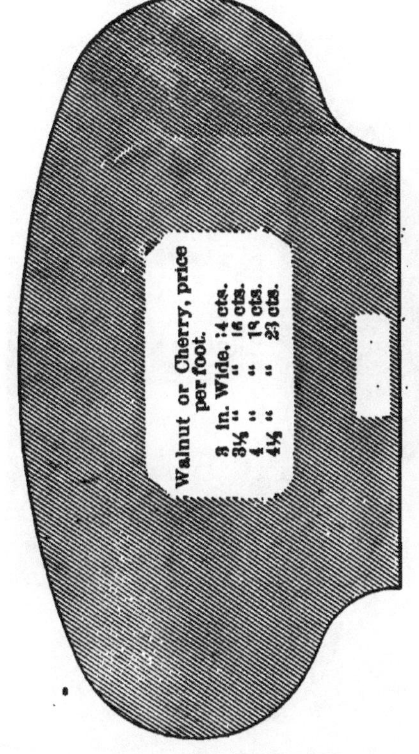

Pine Rail half price of Walnut or Cherry.

No. 751.

Walnut or Cherry, price per foot.

3 in. wide, 14 cts.
3½ " " 16 "
4 " " 23 "
4½ " " 28 "
5 " " 30 "
5½ " " 34 "

No. 752.

Pine Rail half price of Walnut or Cherry.

759

Balusters.

SIZES.	Pine.	Oak, Ash, Butternut	Walnut or Cherry.
No. 782, 1¾ inch	7c	9c	11c
" 783, 1¾ "	9	11	13
" " 2 "		13	15
" " 2¼ "		15	17
" " 2½ "		18	21
" 784, 785, 787, and 788, 1¾ inch		15	16
" " " " 2 "		17	18
" " " " 2¼ "		19	21
" " " " 2½ "		22	24

We turn all Balusters 2 ft. 4 in. and 2 ft. 8 in. long and keep these lengths always in stock. Odd lengths cost extra.

Newels

		Pine.	Oak, Ash, Butternut	Walnut or Cherry.
P. 72	Platform Newel No. 1, 5 in.		$2.75	$5.00
	" " No. 2, 5 "		3.50	6.00
	" " No 3, 5 "		2.50	4.75
P. 68	Newel No. 761, 4 inch	$ .75	1.00	1.50
	" No. 761, 5 "	1.25	1.75	2.25
	" No. 761, 6 "	1.50	2.25	3.00
	" No. 762, 6 "	1.50	2.25	3.00
P. 69	" No. 5, 6 "		6.00	7.50
	" No. 763, 7 "		5.00	7.00
P. 71	" No. 7, 7 "		15.00	18.00
	" No. 8, 8 "		12.00	14.00

For Prices of Stair Rail see Cuts.

Church Seats, Pew Ends, Etc.

Owing to an increased demand for Church Finish, we are giving this department our special attention, and are better prepared than ever before to furnish promptly, made of *thoroughly seasoned material*, Pew Ends and Church Seats of every description, ready to put together, with Cherry, Black Walnut, or Mahogany Back Roll and Partition Cap complete; also Pulpits, veneered or plain.

Mantels, Hardwood Finish, Etc.

We are prepared to furnish of Pine or any of the Hard Woods, veneered or plain, any design or style of Wood Mantels; also, all kinds of Hardwood Finish, Flooring, etc.

Mouldings.

A large and complete line always on hand. Send for Moulding Book.

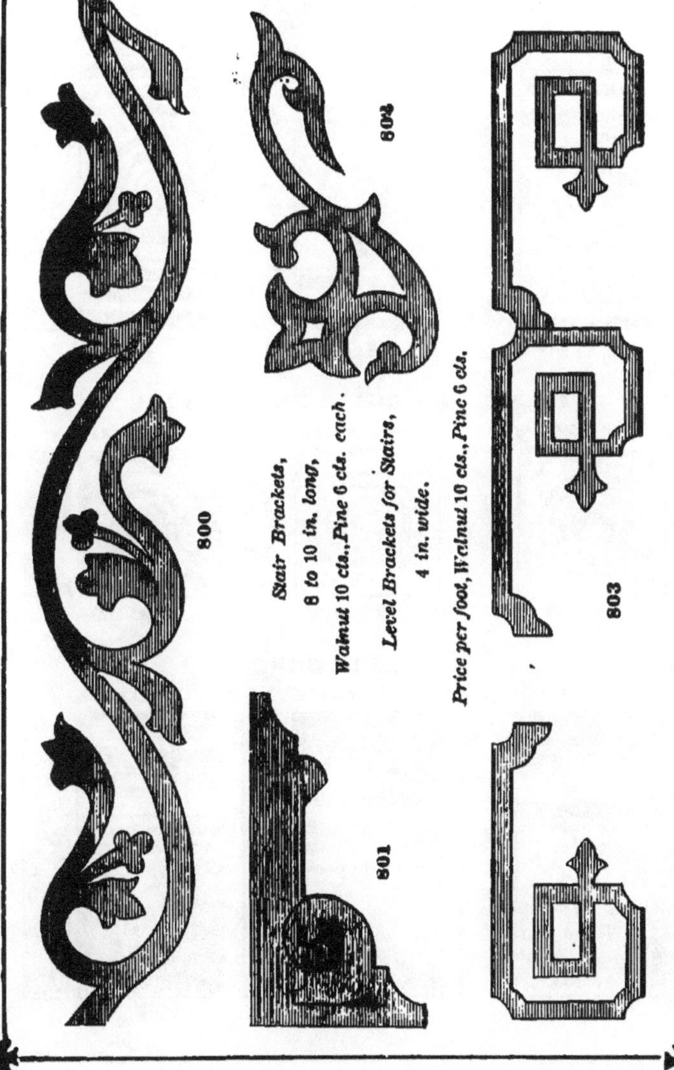
800
802
801
Stair Brackets,
8 to 10 in. long,
Walnut 10 cts., Pine 6 cts. each.
Level Brackets for Stairs,
4 in. wide.
Price per foot, Walnut 10 cts., Pine 6 cts.
803

HEAD BLOCKS.

5½x10x1⅛ 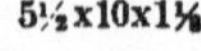 

Fig. 199

Fig. 203

Fig. 200

CORNER OR CHAIR RAIL BLOCKS.

5½x5½x1⅛ | 5½x5½x1⅛ | 5½x5½x1⅛

Fig. 99

Fig. 105

Fig. 100

BASE BLOCKS.

5½x11x1⅜ | 5½x11x1⅜ | 5½x11x1⅜

Fig. 36

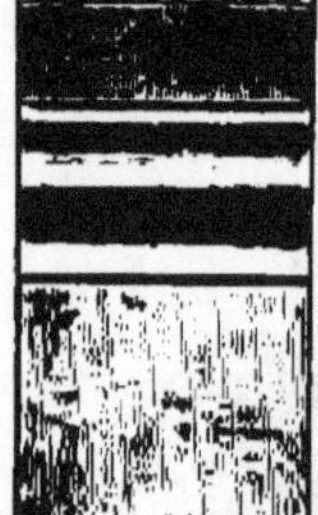

Fig. 37

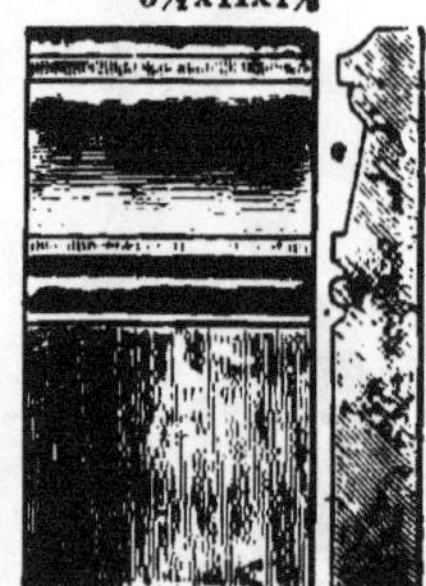

Fig. 6

HEAD BLOCKS.

5½x10x1⅛ | 5½x11x1⅛ | 5½x12

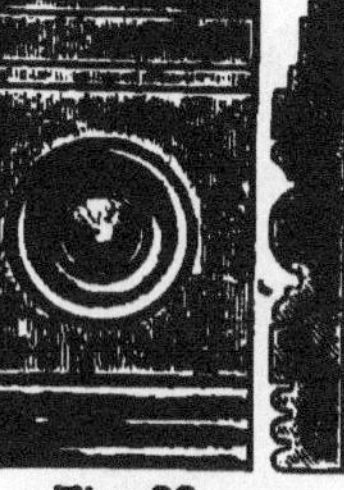

Fig. 32 Fig. 33 Fig. 35

CORNER OR CHAIR RAIL BLOCKS.

5½x5½x1⅛ | 5½x5½x1⅛ | 5½x5½x1⅛.

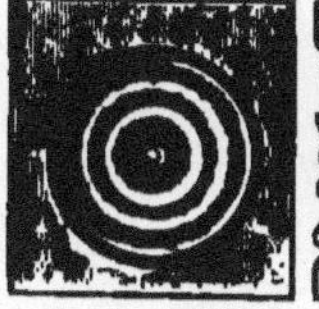

Fig. 107 Fig. 102 Fig. 103

BASE BLOCKS.

5½x11x1⅜ | 5½x11x1⅜ | 5½x11x1⅜

Fig. 9 Fig. 2 Fig. 1

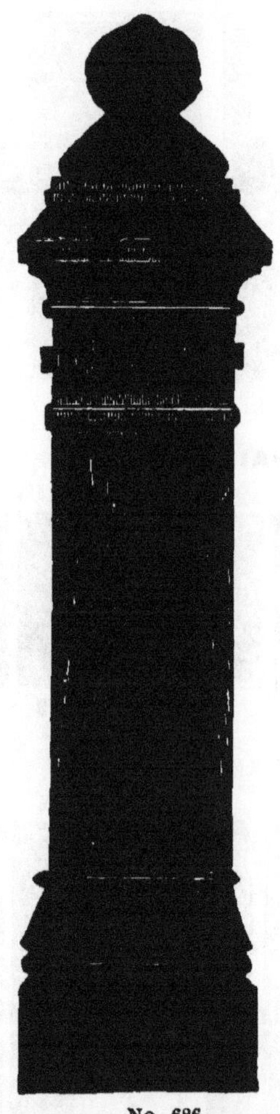

No. 686. **Prices for Pine.** **No. 690.**
No. 690—4x4—2 feet 6 in. high, $1.00 No. 690—6x6—2 feet 6 in. high. $1.25
No. 686—6x6—2 feet 6 in. high, $2.50.

BALUSTERS FOR OUTSIDE BALUSTRADES.

683

684

685

687

688

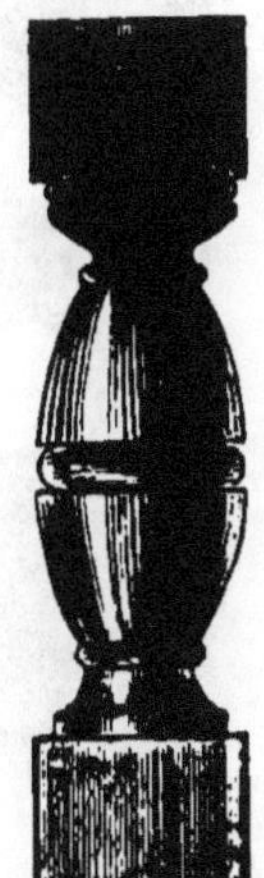
689

PRICES FOR PINE OR WHITEWOOD BALUSTERS.

3x3 14 inches	10 cents each	
" 16 "	11	"
" 18 "	12	"
" 20 "	13	"
4x4 14 inches	14	"

4x4 16 inches	15 cents each	
" 18 "	16	"
" 20 "	17	"
" 22 "	19	"
" 24 "	21	"

SAWED BALUSTERS FOR BALUSTRADES.

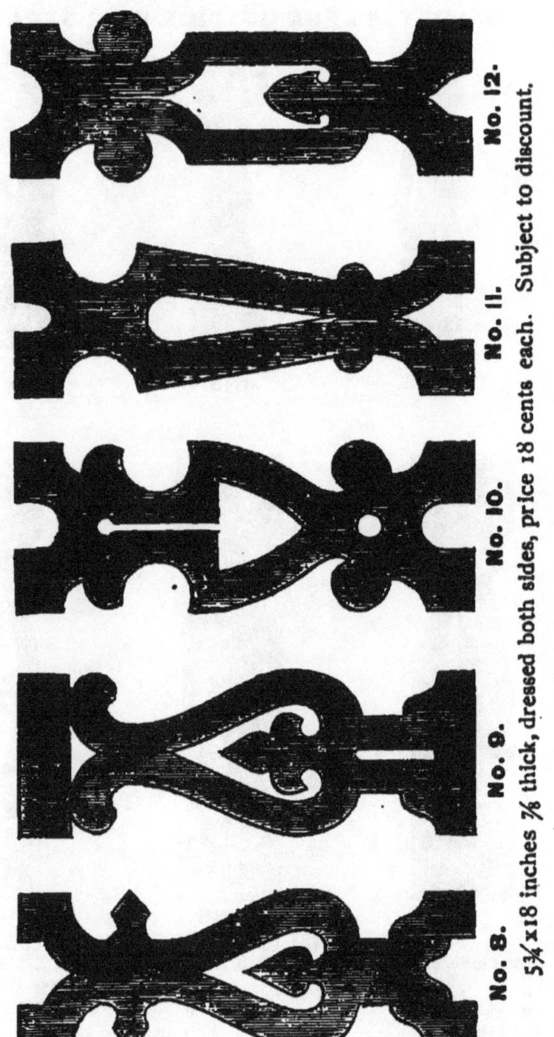

No. 8. No. 9. No. 10. No. 11. No. 12.

5¾ x 18 inches ⅞ thick, dressed both sides, price 18 cents each. Subject to discount.

CRESTINGS.

No. 601.

No. 611.

1⅛ x 8 inches, Pine, 15 cents per foot.

Pine Rail and Base for Sawed Balusters.

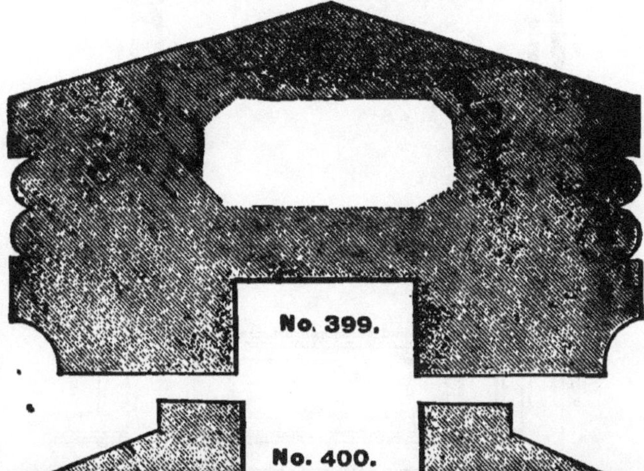

No. 399, price.............................. ....$6.00 per 100 feet.
No. 400, price.............................. 2 70 per 100 feet

VERANDA.

No. 702

Prices for Turned Porch work furnished upon application.

BRACKETS.

Having made a specialty of Brackets and Scroll Work, we are prepared to furnish same promptly of any pattern or design. See prices below:

ONE MEMBER BRACKETS—1¾ inches thick.

Sizes.		
10 x 16,	Nos 621, 622, page 89	$ .30
12 x 18,	" " "	.40
12 x 24,	" " "	.50
10 x 16,	No. 653, page 88	.75
12 x 18,	" " "	.85
12 x 24,	" " "	1.00

ONE MEMBER BRACKETS—1¾ Carved.

7 x 9,	Nos. 656, 664, pages 86, 88	$ .75
10 x 16,	" "	.90
12 x 18,	" "	1.00
20 x 30,	" "	1.50

THREE MEMBER BRACKETS.

7 x 9,	Nos. 613, 615, 617, 618, pages 86, 87	$ .40
10 x 16,	" " "	.50
12 x 18,	" " "	.60
20 x 30,	" " "	1.00
7 x 9,	Nos. 642, 643, 659, pages 86, 88	.50
10 x 16,	" " "	.65
12 x 18,	" " "	.75
20 x 30,	" " "	1.40
26 x 32,	" " "	2.50

VERANDA SCROLLS.

⅞ INCH THICK.

10 x 16,	Nos. 638, 639, page 87	.30
12 x 18,	" "	.40
12 x 24,	" "	.50

CABLES.

Nos. 619 and 620, page 89, according to size.........$6.00 to 10.00

For additional designs see our Moulding Book, which will be furnished free of charge. All above prices for Brackets unmoulded. Above prices subject to special discount.

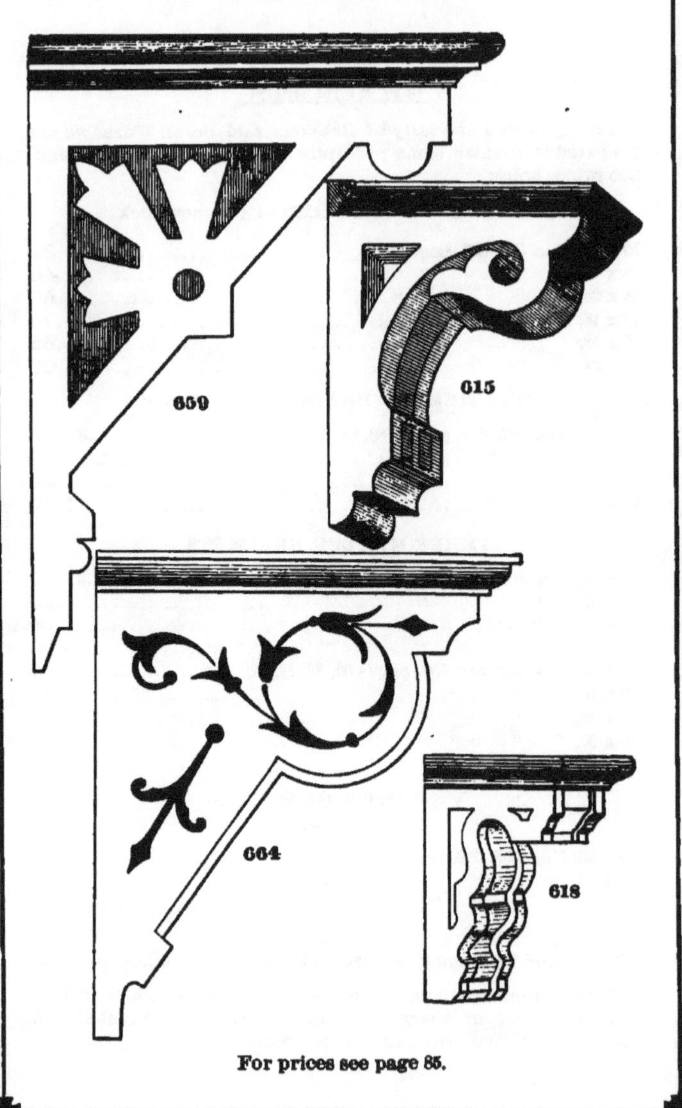

For prices see page 85.

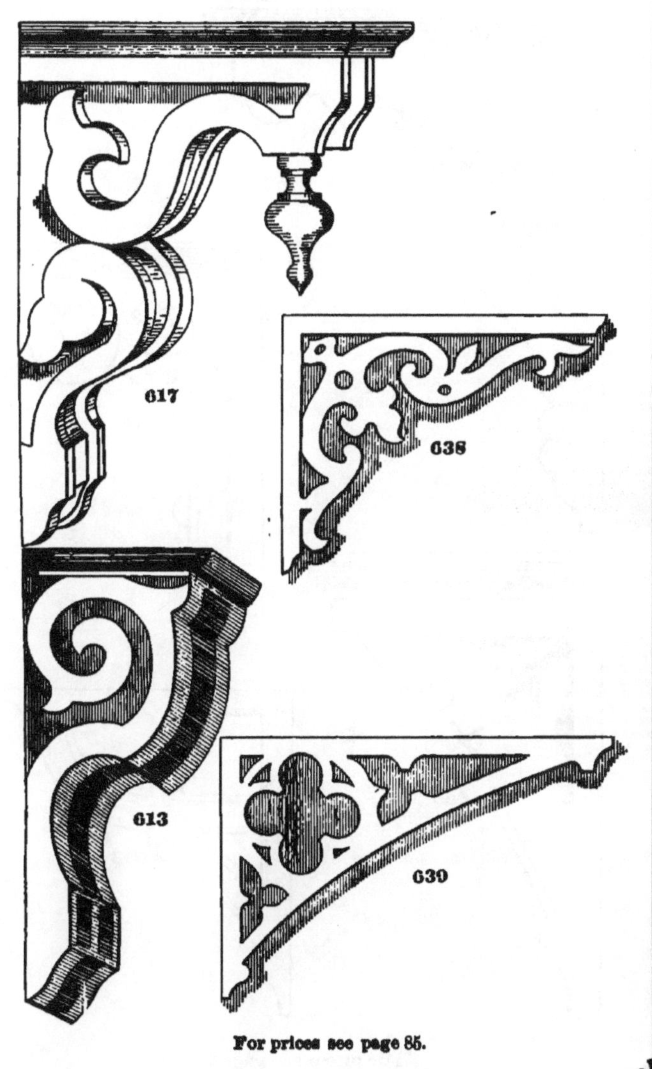

For prices see page 85.

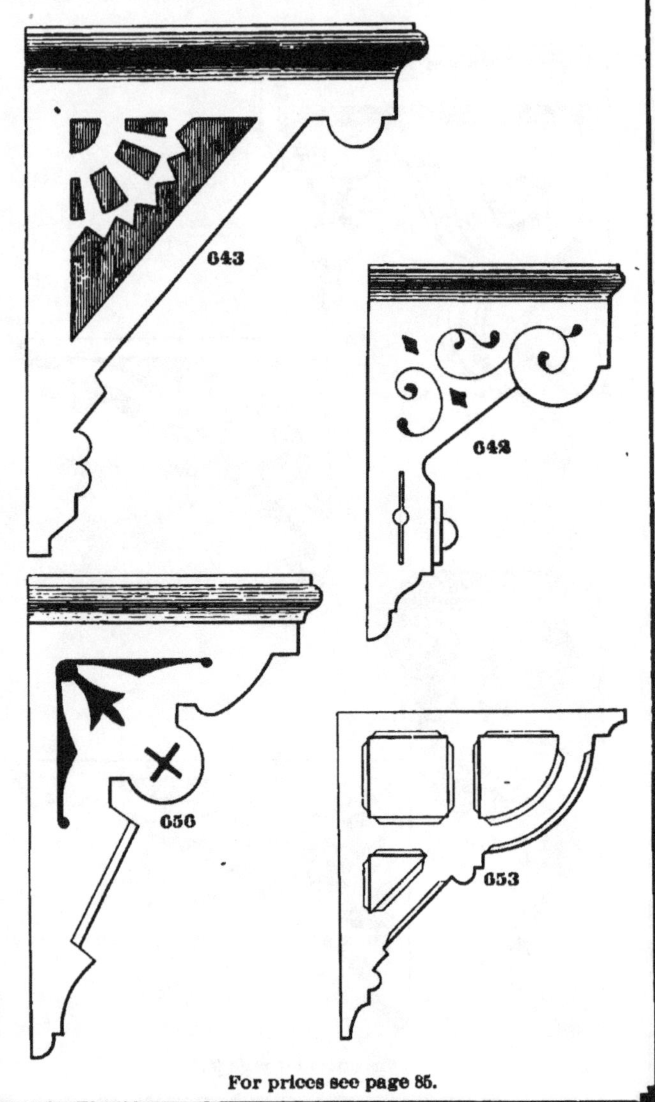

For prices see page 85.

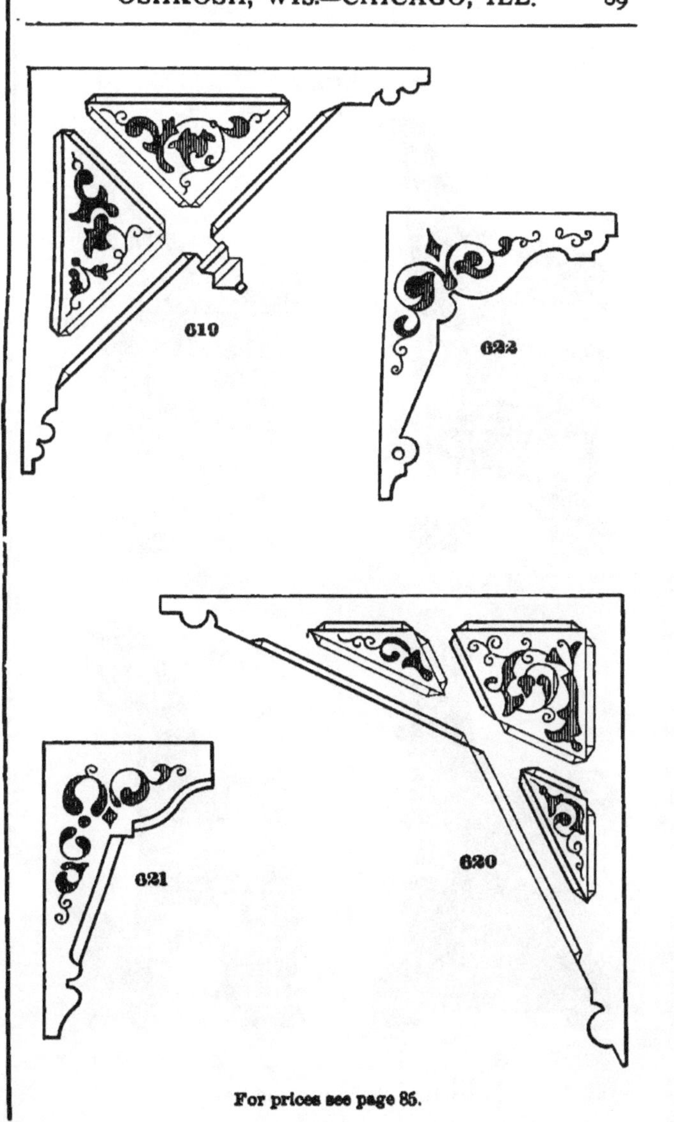

For prices see page 85.

PEW ENDS.

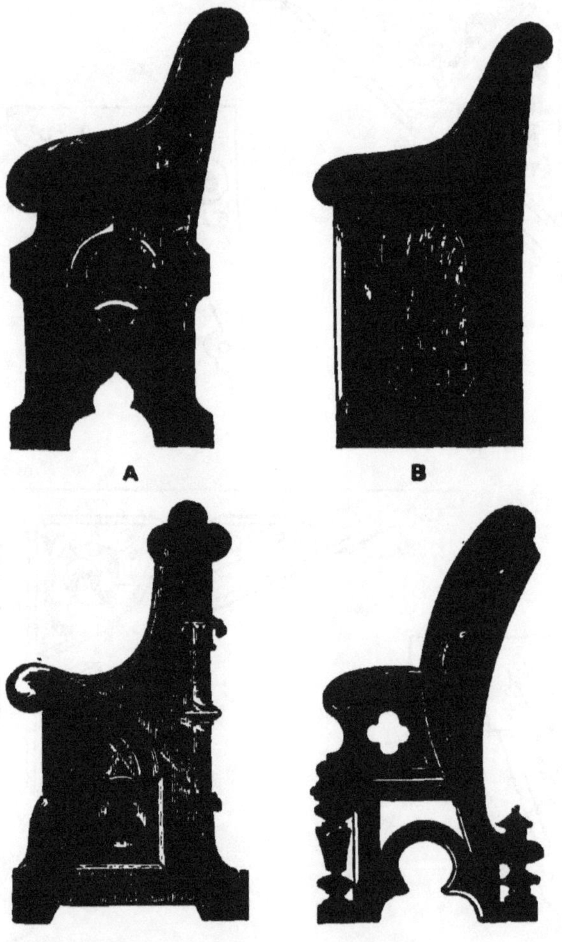

Prices furnished upon application.

PULPIT.

No. 710.

CORNER BEADS.

PRICE-LIST PER HUNDRED.

Four feet in length, one and one-eighth inches in diameter,	$16.00
Four feet in length, one and three-eighth inches in diameter,	18.00
Four feet in length, one and three-quarter inches in diameter,	22.00

Extra lengths and sizes, extra price.

WOOD MANTELS.

719

722

Mantels made of Pine, Oak, Cherry, Walnut or Mahogany.

PRICE OF ENAMELED GLASS.

Description.	Single Thick, per square foot.	Double Thick, per square foot.
Plain per Square Foot	30 cents.	35 cents.
Obscure " " "	35 "	40 "

PRICE OF CUT GLASS.

(Cut by Sand Blast Process.)

We give cuts of the following patterns on pages 94 to 99.

No. of Pattern.	Price per Sq. Foot.	No. of Pattern.	Price per Sq. Foot.
18	$1.25	110	$1.25
41	.75	116	1.25
43	1.75	118	1.50
44	1.75	119	2.00
73	1.00	121	1.50
78	2.00	127	2.00
92	1.50	130	1.50
93	2.00	701	1.25
95	1.75	712	1.25
97	1.50	715	1.75
98	1.50	717	1.50
99	1.25	720	1.75
101	1.25	726	1.50
105	2 00	728	1.50
106	2.00	741	1.50
109	1.00	743	2.00

Above prices *are net and include glazing* and are to be added to the net prices of open Sash Doors. We are prepared to furnish Ornamental Glass with name or initials cut in same.

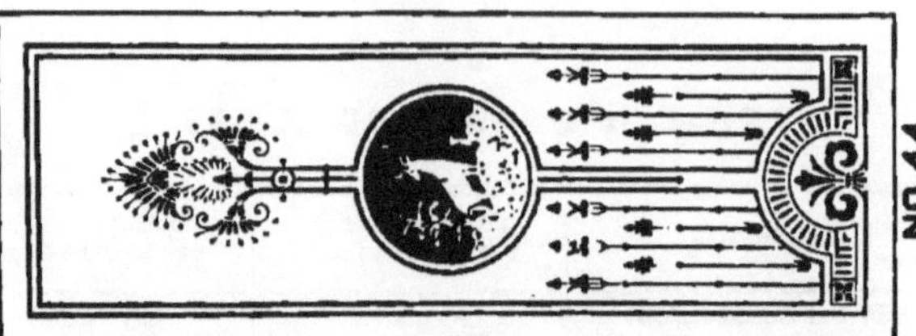

PATTERNS OF CUT GLASS.

For Prices see page 93.

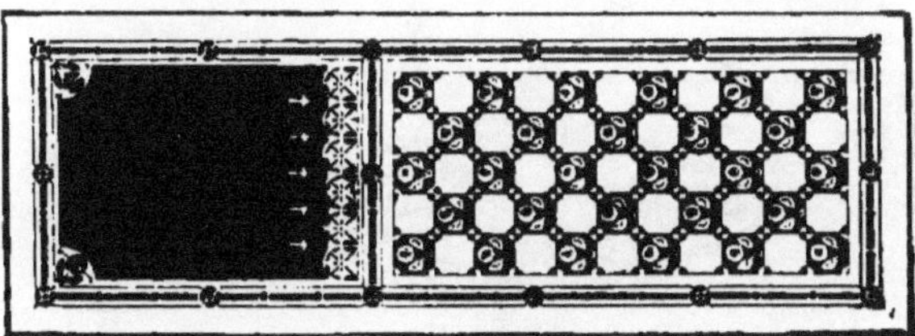

PATTERNS OF CUT GLASS.

For prices see page 93.

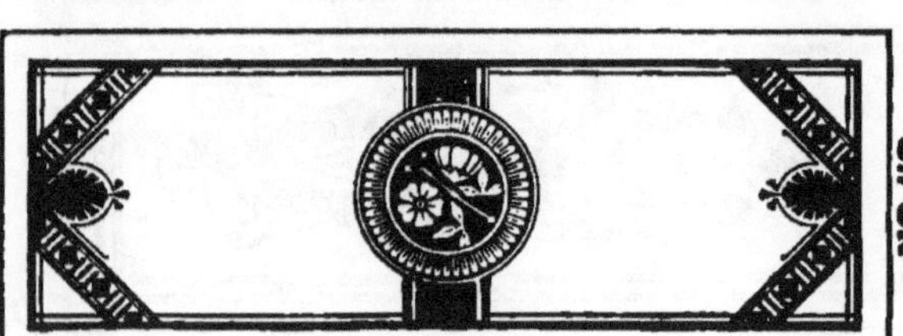

PATTERNS OF CUT GLASS.

For prices see page 93.

PATTERNS OF CUT GLASS.

For prices see page 93.

PATTERNS OF CUT GLASS.

For prices see page 93.

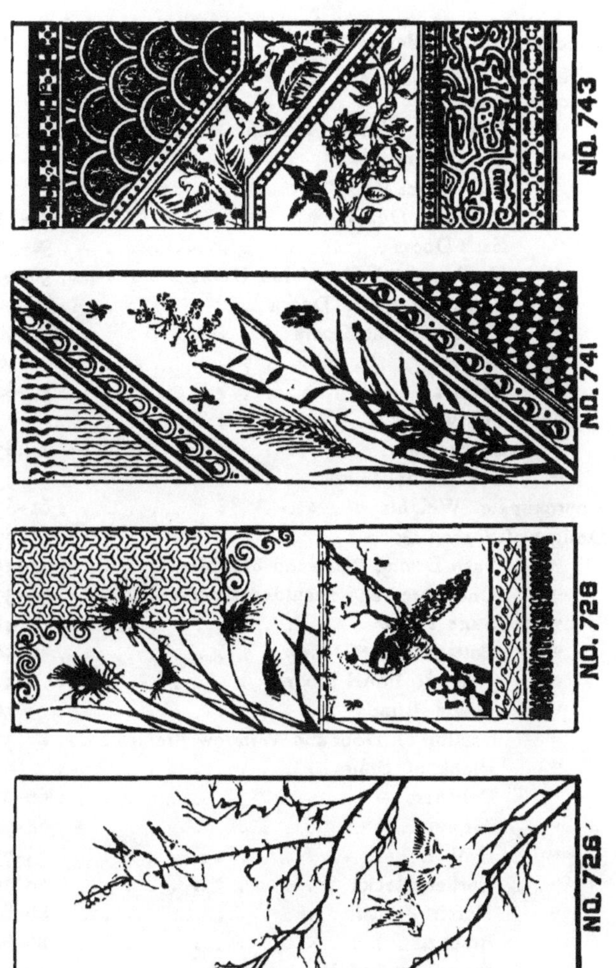

PATTERNS OF CUT GLASS.

For prices see page 93.

CONTENTS.

www.ingramcontent.com/pod-product-compliance
Lightning Source LLC
Chambersburg PA
CBHW021829190326
41518CB00007B/788

* 9 7 8 3 7 4 4 6 5 0 9 7 7 *